FOOD IRRADIATION : THE FACTS

All the information you need to make an informed decision
about what happens to the food you eat.

FOOD
IRRADIATION
THE FACTS

TONY WEBB AND DR TIM LANG
of the London Food Commission

THORSONS PUBLISHING GROUP
Wellingborough, Northamptonshire

Rochester, Vermont

First published 1987

British Library Cataloguing in Publication Data

Webb, Tony
Food irradiation: the facts.
1. Radiation preservation of food
I. Title   II. Lang, Tim
III. London Food Commission
664.0288   TP371.8

ISBN 0-7225-1442-5

Printed and bound in Great Britain

Contents

List of Tables
and Figures

Acknowledgements

The authors are indebted to the members of the Food Irradiation Working Group, staff and Council of the London Food Commission; to the many independent researchers and writers whose work has contributed to our understanding of these issues; and to the many others who helped with the production of this book and the reports and papers that preceded it.No work is ever produced by its authors alone and this has been no exception. Those who have contributed are too numerous to name individually but our special thanks go to Frank Cook MP, Ken Bell, Kitty Tucker, Sue Dibb, Melanie Hare, Angela Henderson, Liz Castledine and Claire-Marie Fortin.

Glossary

ACINF	Advisory Committee on Irradiated and Novel Foods
BEUC	*Bureau European des Unions de Consommateurs*
COMA	Committee on Medical Aspects of Food Policy (DHSS)
DHSS	Department of Health and Social Security
EEC	European Economic Community
EHO	Environmental Health Officer
FAO	Food and Agriculture Organisation (of the United Nations)
FDA	Food and Drug Administration (USA)
FDF	Food and Drink Federation
FIRA	Food Industries Research Association (at Leatherhead, Surrey)
Gy	Gray — unit of received dose of radiation
IAEA	International Atomic Energy Authority
IBT	Industrial Bio-Test Limited
ICRP	International Commission on Radiological Protection
JECFI	Joint Expert Committee of the IAEA/WHO/FAO
kGy	kiloGray
LFC	London Food Commission
MEP	Member of the European Parliament
MP	Member of Parliament
mSv	milli Sievert
NACNE	National Advisory Committee for Nutrition Education
PA	Public Analysts
PUFA	Polyunsaturated Fatty Acid

Rad	old unit of received dose of radiation
rem	old unit of dose measuring biological damage
Sv	Sievert — unit of dose measuring biological damage done in living tissue
TSO	Trading Standards Officer
UK	United Kingdom
UN	United Nations
US	United States
USDA	United States Department of Agriculture
WHO	World Health Organisation (of the United Nations)

Chronology

1916 Sweden experiments with irradiation of strawberries.

1921 Patents taken out in USA.

1930 Patents taken out in France.

1953 Food irradiation to be one of the 'atoms for peace' technologies. US Army begins research.

1957 Irradiation used on spices in West Germany.

1958 Food irradiation banned in West Germany.
      USSR permits irradiation of potatoes.

1960 Canada permits irradiation of potatoes.

1963 USA permits irradiation of wheat, potatoes and bacon.

1968 USFDA withdraws permit for bacon.
      US army studies found to indicate adverse effects and to have been poorly conducted.

1970s Research programme taken over by IBT Ltd.
      IAEA organises 'expert seminars' and publishes reports on food irradiation.
      IAEA sets up joint expert committee with WHO & FAO (JECFI).

1976 JECFI relaxes requirement for testing of irradiated foods so that radiolytic products do not have to pass tests normally required for food additives. Further permits for foods given by various countries.

1981 JECFI gives general clearance up to 10 kGy (average dose) and removes requirement for control of maximum and minimum doses.
      UN Codex Alimentary Process initiated.

Permits extended by various countries.

1982   UK Government Advisory Committee (ACINF) set up to review evidence on safety and wholesomeness.

1983   IBT directors convicted of doing fraudulent research for government and industry. US loses £4 million and 6 years worth of research data.

1986   Scandal over abuse of irradiation by British food companies using Dutch and Belgian irradiators to conceal bacterial contamination on illegal imports to UK and Sweden.

US approves clearance of irradiation for fruit and vegetables up to 1 kGy (and 30 kGy for spices).

UK ACINF report published — recommends there are no special safety problems from irradiation of food up to 10 kGy.

Introduction

This book has grown out of our work at the LONDON FOOD COMMISSION, an independent body set up with help from the now abolished Greater London Council to provide independent research, advice and information on food and food policy issues. However, for both of us, it began many years before this. In the one case out of concern about radiation and health, and in the other, concern about developments within the food industry. These came together in 1985 as a result of an initiative by Frank Cook MP that brought together a number of people concerned about the issue of food irradiation. The result was a working group at the London Food Commission exploring various issues of concern.

Those who control our food supply can easily forget that, as well as the 'experts', there are many people whose interests are rarely considered let alone consulted on matters of food policy. The working group aimed to bring together a wide range of such interests; scientists, public health agencies, consumers organisations, trades unions and environmental groups.

At that time in 1985, we were anticipating the early publication of a report by the UK Government's Advisory Committee on Irradiated and Novel Foods (ACINF) that had been looking into the issue since May 1982. In the event the report was delayed until the spring of 1986. It was widely expected that the ACINF would recommend relaxing the ban on irradiation of food in the UK.

The LFC working group became aware of considerable economic and political pressure for removing the current ban on food irradiation notably that coming from a working group at the Food Industries Research Association at Leatherhead (FIRA). This included representatives of the FIRA, Unilever plc — the multinational British/Dutch food Company, and two firms which stood to gain directly from any change in the law; ISOTRON plc and Radiation Dynamics Ltd. Leading spokespeople from this group were prominently promoting the technology with the media and one was represented as a technical advisor on the government committee.

The publicity line emerging was that there were 'absolutely no problems whatsoever' with food irradiation. That this went unchallenged, indicated that there was considerable ignorance about the issue within the media, consumer bodies, public health agencies, sectors of the industry who should have known otherwise and the public at large.

In order to alert the public and other concerned groups to the possibility of an imminent change in the law, the LFC published a briefing paper in June 1985, a leaflet in July and a detailed report *Food Irradiation in Britain?* in September 1985. These publications set out a number of the issues of concern about food irradiation - issues that need to be discussed as part of any process of public consultation on proposals to lift the current ban on irradiation of food in the UK. It was, we felt, important that the public should be involved in the debate about the introduction of food irradiation technology at the earliest stages. The decisions should not be made by 'experts': rather the experts need to make available the facts on which the public can make an informed decision.

The London Food Commission working group then undertook a pilot survey to explore the policies and attitudes of several sectors of the food industry, and of the major consumer organisations. We felt it was important to find out the views of the food industry rather than assume, on the basis of press reports, that all food firms were pressing for irradiation. A similar survey went out to Trading Standards Officers. The specific aims of these surveys were threefold:

- by asking questions, to stimulate discussion and debate about the possible risks as well as the potential benefits of irradiation;
- to explore the extent to which various sectors of the industry were planning to introduce irradiated food products and the

extent to which they recognised and supported the need for safeguards if and when it was introduced;

● to assess the extent of awareness of actual and potential abuses of irradiation that, at the time, we suspected were going on, and the availability of methods for controlling these abuses.

The concerns of the working group were, to our surprise, echoed by many of those responding to the surveys. The findings indicated that, contrary to the impressions being given in the media, the food industry and consumers were far from united in wanting food irradiation. Most consumers' organisations were highly critical of irradiation and did not think the current ban should be removed without stringent safeguards. While many large companies either declined to participate or indicated they were undecided, a number of medium sized food manufacturers expressed either their opposition, or their concern over safeguards. Even the large companies responding indicated that additional regulations would be needed alongside any change in the law on irradiation.

Retailers, in particular, were clearly worried about consumer reaction and undecided about the benefits of irradiation. Two major retailing companies have since declared that irradiation has no place within their policy of responding to consumer demands for healthier foods and providing quality fresh foods by ensuring rapid turnover of their stock. The British Frozen Food Federation has called for European-wide action to stamp out abuse of irradiation, and recommended that food irradiation should not be legalised until tests are available to detect it. A significant number of Trading Standards Officers indicated concern over the lack of preparation within the monitoring agencies for any legalisation of irradiation.

The working group had been approached by a leading importer of prawns, who was concerned about the illegal use of irradiation which he saw as undermining food hygiene standards in the sea-food and other food processing industries. In April 1985, two days before the publication of the report of the Government Advisory Committee, the Thames TV Programme '4 What It's Worth' revealed publicly the extent of these abuses and named major British companies using irradiation plants in the Netherlands and

Belgium to conceal bacterial contamination in consignments of seafoods.

On the day before the publication of the Advisory Committee report, backbench Members of Parliament tabled three Motions in the House of Commons. These covered: the details of illegal importation of irradiated foods; concerns about the safety and wholesomeness of irradiated foods; and a motion on the conflict of interest of Frank Ley, the director of ISOTRON plc. This last noted Ley's leaking of the main conclusions of the report, the incompatibility of his high profile role as media spokesperson with his position on the advisory committee, the flotation of ISOTRON on the Stock Exchange during the period in question, and the rise in its share values when the financial press linked the future of the company with the impending recommendations of the Committee.

The Committee report itself has been described by some as a travesty of scholarship. Leading Opposition spokespeople in Parliament have called on the government to provide basic scientific references to support the report's conclusions. Major issues of concern were either ignored or inadequately addressed and the public was then given $3\frac{1}{2}$ months only to comment on the recommendations — in essence that there were no reasons to maintain the current ban on irradiated foods.

The details of the pilot survey and the London Food Commission response to the ACINF report were published in a report *Food Irradiation — Who Wants It?* by the London Food Commission in June 1986.

As we go to press the debate about whether the ban on food irradiation should be lifted continues. The Government was forced to allow a further three months for public comment but refused to provide the references requested. There are moves within the European Commission that would undermine even the reluctant assurances on the labelling of irradiated foods that were given in the UK Government's Advisory Committee report.

Information is coming in from the United States on infringements of licences by operators of Irradiation plants and the growing campaign of opposition there. We are in touch with groups and individuals who are similarly trying to halt moves to legalise irradiation in Australia, New Zealand, Malaysia, Germany, Sweden, Norway, and Finland, and with groups in The Netherlands, Belgium and the USA who are becoming aware of the implications of their countries' premature decision to permit its use.

Initially the London Food Commission did not oppose irradiation in principle. We wanted to see a wider debate about all the risks and benefits and for no hasty change in the law before all issues of concern had been addressed. Now, however, we have reluctantly been forced into opposing the introduction of food irradiation because those with vested interests in its introduction have chosen to ignore the issues of concern, and the decision makers have hardly begun to consider them. They have instead chosen to try to win a simplistic victory by playing on public ignorance and using their influence within the machinery of government in order to achieve a rapid change in the law.

This book is an attempt to bring together the facts we have assembled about food irradiation; the causes for concern as well as the potential benefits; the scientific issues and those that apply to the real world; and the pressure points for those who, like ourselves, are concerned to intervene in the public debate about whether, when and under what conditions it should be used.

TONY WEBB & TIM LANG
September 1986

1

The Best Thing Since Sliced White Bread?

The food industry has used a variety of methods over the years to preserve or extend the shelf life of food. These have included cooking, packaging, smoking, chilling, freezing, dehydrating, and using chemical additives.

The latest method now being seriously considered, is the use of ionising radiation. Though some 30 countries worldwide permit irradiation for a variety of different foods, the process is currently banned in Britain and several other European countries. It is illegal in Britain to irradiate food for public consumption or to import irradiated foods. Exceptions have been made for sterile diets for some transplant patients, and for animal feed.

All this could soon change. There are considerable pressures to relax the current ban coming from some sections, though by no means all, of the food industry, and from sections of the nuclear industry that see food irradiation as an extension of the 'atoms for peace programme' or as a way of utilising nuclear waste.

It is claimed that irradiation of food can be used to inhibit the sprouting of vegetables, delay the ripening of fruits, kill insect pests in fruit, grains or spices, reduce or eliminate the micro-organisms that cause food to spoil and in particular reduce the bacteria on some meats and sea food products that cause food poisoning. It has been hailed as an alternative to other methods of preservation such as the use of chemical additives. It is claimed that the process is completely safe, and that consumers will benefit from reduced wastage, greater convenience and better quality food.

Against this view, there is a body of opinion that points to a variety of adverse effects from irradiation; such as unique chemical changes, loss of vitamins, off-flavours and smells, a limited range of foods for which it has been found suitable, the necessity for use of additives to offset undesirable effects, and studies showing adverse health effects in animals and humans fed on irradiated foods.

Concern has been expressed by several trades unions over the possible health effects on members working in irradiation plants; and by environmental groups, over the transport of radioactive material and discharge of radioactive effluent at local community level. Such critics point to violations of worker and public safety regulations that have occured in irradiation plants in the USA, and to the inadequacy of the current UK regulations on radiation protection.

Consumer groups have expressed concern over labelling of irradiated foods. At the very least they believe that the consumer has a right to know if any food, or ingredient of processed foods, has been irradiated. Irradiated foods will look fresh and retain the appearance of freshness longer. Without labelling the possibility of counterfeit 'freshness' can be easily exploited.

Such concerns are reinforced by evidence of significant nutritional losses in irradiated foods; especially the severe damage to some vitamins and to essential poly-unsaturated fats and fatty acids. It is now widely recognised, not least by the 1984 report from the Government's Committee on Medical Aspects of Food Policy (COMA) on the impact of diet on heart disease,[95] that there needs to be a change in the British diet. The very foods that the nutritional consensus now recommends; increased consumption of white meats, whole grains, fresh fruits and vegetables; are the very foods targeted for irradiation and in which significant nutrient losses will occur. We, and a growing number of nutritionists are deeply worried that use of irradiation will undermine the recent progress in promoting dietary change. The public are genuinely interested in and concerned about issues of health.

Not all sectors of the food industry want irradiation. Some are worried about the possibility of its introduction, others are concerned about additional regulations in the food trade that may be needed if it is introduced, many are worried about consumer reaction and are hedging their bets. Some have spoken out strongly against it either from self interest or for more principled reasons.

Irradiated food is already coming into Britain, despite the ban. There is also evidence that some food suppliers have used the process to eliminate high bacterial loads on food in order to make them saleable.[101] Besides being illegal, this practice is in violation of recommendations of the World Health Organisation, and could, in some cases, lead to serious health hazards. Irradiation may reduce the bacterial load on food. It does not eliminate the chemical toxins that may have been created by earlier contamination.

The Government has refused to act on documented cases of abuse, saying that responsibility rests with local authorities.[91] At the present time there is no test that can be used by port health and trading standards officers to detect irradiation. It will be several years before a suitable test to detect irradiation in all food is developed and longer before tests are developed that will enable us to tell if food has been re-irradiated and, if so, how many times and with what dose(s).

There is considerable emotional reaction to the term 'irradiation'. Some sectors of the industry fear that this will lead to unreasonable consumer resistance to irradiated foods. In some countries there have been moves to ban the use of the term 'irradiated' and to substitute a less emotive symbol or word for labelling purposes. Consumer organisations in the UK are united in their insistence that all food must be clearly labelled as 'irradiated' — thus ensuring that consumers get the choice between irradiated and non-irradiated foods. The labelling of loose fruit and vegetables and of food sold in restaurants will pose enormous enforcement problems.

Look at it this way. If you were a market stall holder trying to sell irradiated fruit and veg and your customers wouldn't buy it, would you label it knowing that no Trading Standards Officer (TSO) could prove it was irradiated? Or, put another way, if irradiation was widely accepted and welcomed by consumers, what is there to stop you claiming that the food you are selling has been irradiated? The whole system is wide open to counterfeiting based on the way irradiated foods appear fresh for longer.

The facts about food irradiation go beyond the pure sciences of toxicology, microbiology, and nutrition. They extend to issues of concern about irradiation in the real world. In the real world of food supply and consumption, money and cost influence decisions about what is acceptable. Nowhere has there been an official statement about what financial benefits are to be had and, more importantly, by whom. It hardly inspires public confidence to discover that some who stand to gain most have been intimately involved in the Government's decision-making process, and, at the same time have been promoting the idea of food irradiation and the interests of their own companies; which have then benefited from speculation that the ban on irradiation is about to be removed.[102]

If the current ban on irradiation of food is to be removed then we need to ensure that it will bring real benefits to consumers. We need to be sure that there are either no risks or that risks to the public and to those who work in the food industries are reduced to a minimum by effective regulations. We will also need an effective system for enforcing these regulations that can prevent abuses.

A Brief History of Food Irradiation

The idea of irradiating food is not new. The treatment was tested on strawberries in Sweden in 1916. The first patents on the idea were taken out in the US in 1921, and in France in 1930. Little progress was made until 1953 when US President Eisenhower announced the 'Atoms for Peace Programme'. Public attention was to be shifted away from nuclear weapons by promoting nuclear power and other uses of nuclear technology, so that the academic and industrial infrastructure could be developed behind which the weapons programmme would continue. There followed a decade of intensive research into food irradiation funded and supervised by the US Department of Defence.[1,2]

The first commercial use of food irradiation actually occurred in Germany in 1957 for sterilisation of spices used in sausage manufacture. This was brought to an abrupt end when the German government banned the process in 1958. The Soviet Union was the first government to permit irradiation; for inhibiting sprouting of potatoes in 1958, and disinfestation of grain in 1959. Canada permitted its use for potatoes in 1960. The US Food, Drug and Cosmetics Act of 1958 defined the irradiation process as an additive. Users have to petition the US Food and Drug Administration for permission to market irradiated products. This has resulted in stringent requirements for testing of irradiated foods in the US. It was not until 1963 that clearance was given for sterilisation of can-packed bacon and the inhibition of potato sprouting and wheat disinfestation already in use elsewhere. The clearance on bacon was withdrawn in 1968 after a review of the research found adverse effects and deficiencies in the conduct of the experiments.[29]

Since then around 30 countries have permitted irradiation of 28 different foodstuffs for public consumption. Commercial activities were planned in a further 11 countries as of January 1985.[73]

Britain, along with West Germany, and most of the Scandinavian countries, currently does not permit irradiation of food for public consumption.[7] There does exist the mechanism for government to permit its use under the Food (Control of Irradiation) Regulations 1972[8] subject only to approval of the facilities and the system for recording doses administered to the food. So far these regulations have only been used for irradiation of food for hospital patients needing sterile diets, such as transplant patients, and a few public relations exercises. Some 800 tons of animal feed, mainly for experimental animals has also been irradiated.[1]

This may soon change as a result of the report of the UK Government's Advisory Committee on Irradiated and Novel Foods,[9] and with the growing pressures for harmonisation of national legislation within the countries of the EEC.[10]

The mixture of protective and permissive national regulations clearly acts as a barrier to international trade in irradiated food products which food manufacturers would like to see removed. After a series of seminars organised through the International Atomic Energy Authority, (IAEA)[3] a Joint Expert Committee of the IAEA, the World Health Organisation (WHO), and the Food and Agriculture Organisation (FAO) of the United Nations was set up in 1971. This WHO/IAEA/FAO committee was given the task of reviewing and guiding the research into issues of concern. It has

Table 1: FOODS GIVEN PERMITS FOR IRRADIATION IN OTHER COUNTRIES

Foods	Argentina	Bangladesh	Belgium	Canada	Chile	Denmark	France	Hungary	Israel	Italy	Japan	Netherlands	Norway	Phillippines	Poland	South Africa	Spain	Thailand	Uruguay	USA	USSR
Potatoes	×	×	×	×	×	×	×		×	×	×	×			×	×	×	×	×	×	×
Onions		×	×	×	×		×	×	×	×		×			×	×	×	×			×
Garlic			×				×			×						×					
Shallots			×				×														
Wheat and other grains	×		×	×								×								×	×
Spices	×	×	×			×	×					×	×							×	
Chicken	×		×				×			×		×				×					
Fish	×						×					×									
Frozen shrimps	×											×									
Froglegs	×											×									
Rye bread												×									
Egg powder												×									
Blood proteins												×									
Cocoa beans				×								×									
Dates				×																	
Pulses	×			×																	
Pawpaw	×			×												×					
Mango	×			×												×					
Strawberries			×	×								×				×					
Paprika				×																	
Mango achar																×					
Bananas																×					
Dry food concentrate																					×
Dried fruits																					×
Mushrooms												×									
Endive												×									
Asparagus												×									
Batter mix												×									

produced key reports in 1976[4] and 1980[5] on 'The Wholesomeness of Irradiated Foods'. At the same time The Codex Alimentarius Committee of the United Nations (UN) has formulated International Guidelines on irradiated foods based on the recommendations of this joint committee.[6]

The UK government set up its Advisory Committee on Irradiated and Novel Foods in May 1982. This committee reported in April 1986. Like the IAEA/WHO/FAO committee, it recommended that there were no special safety reasons why food should not be irradiated up to a dose of 10 kGy.

On the other hand the US Food and Drug Administration has recommended that general irradiation only be permitted up to the lower dose of 1 kGy — one-tenth of the Codex level. The exception to this is spices which are to be permitted doses up the 30 kGy.

2

What is Food Irradiation?

Very simply, it is a treatment involving very large doses of ionising radiation to produce some desired changes in food, particularly those allowing longer storage or 'shelf life'. This section covers the nature of radiation and what it does, the types of irradiation plant, radioactive sources used, the doses involved and the question on everyone's mind — can food become radioactive?

Ionising Radiation

Radiation is a household word that covers a wide spectrum of energy. At the low end of the spectrum are the emissions from power lines and visual display units. Higher up are radiowaves and microwaves. Also included are infra-red, visible and ultra-violet light, and at the upper end are found X-rays, and gamma rays from radioactive material.

Figure 1: RADIATION   (Source: Ref. 11)

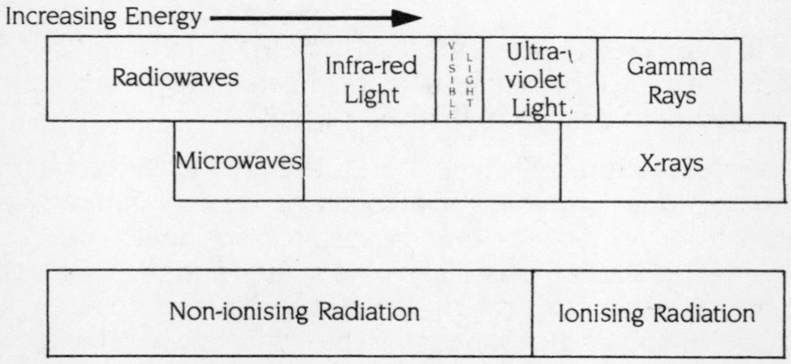

Increasing Energy ———————▶

Radiowaves	Infra-red Light	V I S I B L E L I G H T	Ultra-violet Light	Gamma Rays
	Microwaves			X-rays

Non-ionising Radiation	Ionising Radiation

When radiation strikes other material it transfers its energy. This energy transfer can cause heating, as with microwave cooking, or lying in the sunshine. At a certain level the radiation has sufficient energy to knock electrons out of the atoms of the material bombarded. This can break the molecular structure of the material, leaving positively and negatively charged particles called 'ions' or 'free radicals'. Above this level the radiation is called ionising radiation. The ions are chemically very active and easily recombine or initiate chemical reactions with surrounding material.

Thus ionising radiation alters the chemical structure of material. This in turn can have biological effects on the behaviour of living organisms and the materials they feed on.[12]

Some of the biological effects in irradiation of food can be considered desirable. Irradiation of living organisms, especially people, is almost always damaging.[11,13,14,15]

Food Irradiation Plants

There are currently some 30 planned or operating food irradiation facilities worldwide.[1,73] There are ten plants in Britain which currently irradiate medical supplies or animal feed. Of these only four, all owned by ISOTRON plc, will be able to handle commercial food irradiation. One other may be able to do so and privatisation of some hospital facilities might open possibilties for others on a small scale.[61,72]

In general, an irradiation facility consists of:

● **a loading facility** where the food to be irradiated is packaged and pre-treated by heating and/or refrigeration as needed and loaded onto conveyors that will carry it to:

● **the irradiation cell** where food is exposed to the radiation source. This can be either a cobalt or caesium gamma ray source or an X-ray or electron beam machine source. The distance the food passes from the source and the length of time it is exposed will determine the dose that the food receives. The size of the batch will control the extent to which the food

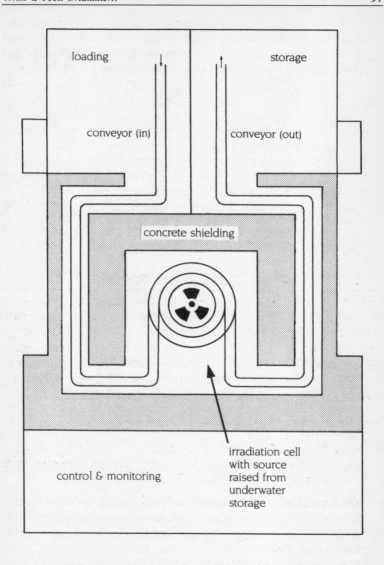

Figure 2: LAYOUT OF AN IRRADIATION PLANT

is uniformly exposed or whether there will be large differences between maximum and minimum doses to different parts of the food batch. Thick concrete shielding protects workers from direct exposure to the source. It then passes into:

- **a storage facility** where it is removed from the conveyor and stored at the required (usually low) temperatures before being despatched to long term storage or retailing outlets.

In addition the plant needs:

- **a fuel handling unit** where irradiated radioactive sources in the form of sealed rods or strips of Cobalt 60 or Caesium 137 are received and loaded by remote handling into the irradiator. They are usually stored underwater and raised from this into the irradiation cell to expose the food.

- **a control unit** governs the movement of the food (and the radioactive sources) through the irradiator.

- there are also facilities for monitoring doses to the food and keeping records.

Since there is no way of detecting that food has been irradiated, or how many times and with what doses, it is vital that strict control is maintained at the source of irradiation and that appropriate certification of the doses given accompanies the food thereafter.

The shape of the irradiation cell, the siting of the radioactive sources and the path of the food through the cell depends on the type of food being irradiated. Some types of irradiator use a central source and pass the food around it on one of a number of circular tracks.

Some foods such as fish may pass along a tube or between two parallel sources. There are proposals for low level irradiators for use on board ships. The plan is to then further irradiate the fish when it comes ashore.

Others may use a simple conveyor belt past either a gamma ray or machine source.

The French have designs for a small scale field irradiator for potatoes, as do others. [123]

Radioactive Food?

The first concern of most people is whether or not food becomes radioactive. Ionising radiation with high energy can cause radioactivity to be created in the material that is bombarded. [15]

The energy level is usually expressed in 'electron volts' (eV). Above about 10 to 15 million electron volts (MeV) it is possible for significant amounts of radioactivity to be created.

It is therefore important that only lower energy ionising radiations are used in irradiation of food.

Even so it is still possible for some trace metal compounds in the food to be made radioactive. Below the 10 MeV level however, the amount of this induced radioactivity is small and decays very rapidly. If foods are stored before use, the level of radiation is likely to be insignificant and well within the range of natural radioactivity found in food anyway. [1]

The other way food could become contaminated with radioactivity is if the radioactive source were damaged. Obviously great care will be taken to prevent this kind of accident. [16]

Thus, provided irradiation is properly controlled, food should *not* become radioactive.

Radioactive Sources for Food Irradiation

It has been suggested that the nuclear wastes from power stations could be used as a radioactive source for food irradiation. [16] This is not immediately feasible because these wastes contain a wide range of radioactive materials some emitting radiation with energy above the critical level. [1]

Two radioactive materials have been identified which are suitable and have energies considered low enough to be safe for use in food irradiation; Cobalt 60, and Caesium 137.

Cobalt 60 gives off two gamma rays of 1.17 and 1.33 MeV and Caesium 137 which gives off 0.66 MeV gamma radiation. A mixture of Caesium 137 and Caesium 134 can also be used. Radiation from these radioactive isotopes is well below the 10 MeV threshold and after storage there should be no measurable levels of induced radioactivity.

Disposal of radioactive cobalt and caesium currently presents a considerable problem because of quantities produced in nuclear wastes from power stations and the length of time they take to decay. The nuclear energy and weapons industries are therefore keen to find commercial uses for them. On the other hand, the viability of using these sources depends on the continuing existence of the nuclear power programme.[16]

The other possible sources for irradiation include beams of electrons, and X-rays which are created when a metal target is bombarded with electrons. These 'machine' sources give off a wide spectrum of energies and greater care must be taken to ensure that the maximum energy is below the 10 MeV threshold. Electrons do not penetrate materials as far as gamma or X-rays and so are only useful for irradiating the surface of foodstuffs.[1]

Radiation Doses

Dose is commonly used to mean one of two distinct things. It is either the amount of radiation received or, in the case of exposure to people, it is related to the amount of biological damage done. In measuring the dose to food we are concerned about the amount of energy that has been deposited as a result of irradiation. The old unit for dose was the Rad; this now being replaced by the Gray (Gy). Very large doses up to one million Rads are being considered for food irradiation. Therefore the doses are more often given in Mega rads (MRads) or Kilo Gray (kGy), a million Rad (1 MRad) = 10 kGy (10,000 Gray).

To give some idea of the scale of these doses, a chest X-ray will deliver about 10 milliRad (0.01 Rad) and an average dose from natural background radiation will give about 100 milliRad (0.1 Rad) per year. The food irradiation dose is therefore 10 to 100 Million times these common doses.

The changes produced in the food increase as the dose goes up. This applies to both the desired and the undesirable changes.[1,2]

Uses of Food Irradiation

As explained above, radiation produces chemical changes in the food and these in turn have biological effects. In some cases the exact mechanisms are still not fully understood.

The following effects can be produced:

Radurisation — low doses, usually below 1 kGy (0.1 MRad)

- sprouting of vegetables such as potatoes and onions can be inhibited so that they keep longer.

- Ripening of fruits can be delayed so that they keep longer and can be transported longer distances.

- Insect pests in grains such as wheat and rice, or in spices and some fruits can be killed. This might replace current methods involving gas storage or fumigation treatments that are hazardous to workers, and could reduce losses of foodstuffs.

Radicidation — medium doses, between 1 kGy and 10 kGy (1.0-MRad)

- The number of micro-organisms that lead to food spoiling such as yeasts, moulds and bacteria could be *reduced* and so extend the life of foods or reduce the risk of food poisoning. This might be important in the case of salmonella in chicken or fish.

Radappertisation — High doses above 10 kGy (1 MRad)

- At these extremely large doses, higher in fact than the 10 kilo-Gray (1 Mega Rad) doses being proposed at present, food can be completely sterilised by killing all bacteria and viruses. This would be used mainly for meat products allowing them to be kept indefinitely.[1]

These effects can be said to extend the 'shelf life' of foods. In these cases irradiation is used as a preservative.

Improvement of Food by Irradiation

In addition to the food preservation benefits claimed for irradiation a number of other 'improvements' in quality have been claimed.[36]

Improvement of baking and cooking quality of wheat including the ability to add up to 15 per cent soya flour to wheat flour without loss of baking quality. Irradiation also 'improves' the elasticity and volume of dough in bread making. A number of additives are currently used to increase the bulk, and the water and air content of the standard white loaf. Yeast can be stimulated by irradiation leading to faster bread making.[2] While this has obvious benefits to the large baking firms it is a matter of opinion whether this leads to an improvement in bread quality.[37,77]

Irradiated barley can increase yield during malting by 7 per cent — a fact of interest to the brewing industry. Irradiation can be used to 'age' spirits[38] and irradiated grapes yield more juice when processed — possibly benefiting the fruit juice and wine making and distillery industries. Irradiated sugar solutions can be used as anti-oxidants, possibly replacing other chemicals used for this purpose in processed and prepared foods.

The time needed to reconstitute and cook dehydrated vegetables, peas and green beans is reduced if these are irradiated. Since cooking times for dehydrated foods are already very short it is debatable whether this is of any real benefit.

It is claimed that irradiation enhances the flavour of carrots and suggested that it could be used for 'tenderising' of meat.[1]

Food that has been contaminated or spoiled can be sterilised by irradiation and so made safe for human consumption. The joint expert committee of the WHO/FAO/IAEA are explicit that irradiation should not be used to make an unsuitable product suitable for consumption, and that food should always be wholesome before irradiation. There are however, pressures arguing that this should be considered and, as we will show, some documented cases of irradiation being used to conceal contamination.[17,101] Most of the basic mechanisms of these 'favourable' changes in food quality are not fully understood.[1,2,12]

On balance most of the above uses are not necessities. They are either luxuries or techniques which might benefit the manufacturer with no clear benefit for the consumer.

Other non-food, but food related, uses include modification of starches for the paper and textile industries and radurisation of gelatin for use in the photofilm industry.

3

Is Food
Irradiation Safe?

In this section we examine the evidence for and against the claims that irradiated food is completely safe.[19,75,76]

The Joint Expert Committee of the International Atomic Energy Authority, The World Health Organisation and The Food and Agriculture Organisation of the United Nations says it is safe.[5] In Britain, the Advisory Committee on Irradiated and Novel Foods says it is safe.[81] Experts from the food, and irradiation industries say it is safe.[19,75,76] Who are we to say it is not?

The answer to that question is to pose another. Are we being asked to take the word of experts or is there conclusive evidence backed up by scientific research that can be referred to and checked by independent researchers?

The Early Research

This is not a matter of being pedantic. Those who have been lobbying for irradiation have suggested that there is 40 years of research showing that there are no safety problems whatsoever. In this we presume they include the first ten years of research done by the United States army in the 1950s and 60s. Research that was used to obtain the clearance for can packed bacon in 1963 — clearance that was subsequently withdrawn in 1986 when the US Food and Drug Administration (FDA) found the data to be flawed. The FDA found significant adverse effects produced in animals fed irradiated food and major deficiencies in the conduct of some experiments. The adverse effects included decreases in surviving weaned young for animals fed irradiated bacon and greater losses of young for those eating bacon exposed to higher doses of radiation.[32]

This same 40 years of research would presumably include also the subsequent research done in the US when the work was turned over to Industrial Bio-Test Limited. Three directors of IBT were convicted in the US courts in 1983 for doing fraudulent research for government and industry. The government uncovered such problems as failure to conduct routine analyses, premature death of thousands of rodents from unsanitary laboratory conditions, faulty record keeping, and supression of unfavourable findings. Prior to the convictions, the Army had declared the beef and pork feeding studies being conducted by IBT in default for similar contract violations. The Army discovered:

> '... missing records, unallowable departures from testing protocol, poor quality work, and incomplete disclosure of information on the progress of the studies.'

The government lost around 4 million dollars and six years worth of animal feeding study data on food irradiation.[105]

Evidence or Opinion?

There are problems involved in testing irradiated foods. Normal tests for chemical safety require isolation of the chemical, and feeding in large quantities to animals on the grounds that, if no problems are found with high doses, then small doses should be safe. While the levels or induced radioactivity in food are so small as not to be an issue of concern, we are concerned that the chemical changes might produce harmful chemicals, especially as some of these are unique to irradiation. Isolating and testing all of these chemical products would be a massive task. It may be that the normal animal testing programme is inappropriate for irradiation. The fact remains that the decision of the IAEA/WHO/FAO expert committee to relax the requirements for testing of irradiated foods in 1976, means that the chemical products of irradiation have not been tested with the same stringency that would be required, say, of chemical additives. We have every reason to be concerned about a number of permitted food additives in use today.[82] Any less stringent testing programme does not inspire confidence.

On the question of the assurances of food safety being based on fact or opinion; the fact is that neither the IAEA/WHO/FAO joint committee nor the UK Government Advisory Committee (ACINF) provide detailed references to the scientific literature to support their conclusions. Presumably if you are an expert in the field and you know of all the research being done, you might be able to infer from the lists of reports and studies that are provided, which evidence is being used to support which claim. If you are not you have no chance of untangling the mystique. You will have to trust (or doubt) the experts. We happen to believe that democratic decision making deserves better than this.

Given the current public mood on many similar issues, the reaction to the suggestion that we should 'trust the experts' is more likely to be doubt than trust and this will damage the chances of irradiation being introduced even in those areas where, on balance, the benefits might outweigh the risks.

There is, in fact, a body of scientific literature indicating adverse effects from feeding of irradiated foods. It may be that we can safely ignore this evidence on the basis of flaws in the data or on the basis of systematic investigation of all the possible effects and potential causes. If so, we respectfully ask to see this evidence. We have already respectfully asked of ACINF, and of the UK Department of Health and Social Security (DHSS) as have back bench MPs and the Opposition spokespeople on Agriculture and Health.[103] All to no avail — the requests have been refused. This is all the more disturbing as we cannot conceive of how the committee could have undertaken its work over the four years without some systematic plan for reviewing each issue of concern and identifying and reviewing the scientific research that provides evidence for and against safety problems in each area. This would have been required of a university undergraduate degree student, let alone a scientific committee advising the Government of an issue of some social importance.

Given that the Committee concluded there were no special safety problems associated with consumption of irradiated food if it is properly controlled, but has not so far seen fit to back up the statements on safety, nor to recommend adequate systems for control, can we be blamed for not finding its conclusions reassuring?

Good to Eat?

We will now consider some of the basic issues about safety of irradiated foods in terms of wholesomeness, vitamin losses, food quality, additives, toxic chemicals and microbiological hazards. These are issues we had hoped the ACINF would have addressed. Later, in Chapter 5, we will take some of these issues up again when we reconsider how the ACINF dealt with them.

The effects of irradiation on food are regarded as desirable from the point of view that they increase the storage time or 'shelf life' of foods. Irradiation should not however be regarded as a panacea for all food preservation problems. Along with the desired effects a number of highly undesirable ones are also produced. Reducing some (but not all) of these may be possible through use of heat or of very low temperatures, removal of oxygen during irradiation, or use of some chemical additives. In these cases irradiation becomes only a part of the preservation process.

In many cases refrigeration will still be needed throughout the storage life of the product and irradiation will not replace this as one of the main methods of preservation. For long term storage sealed packaging may often be needed as recontamination can occur at any time.

Food irradiation should not be seen as a technical solution to all food hygiene problems.[1] Indeed, as we shall show, its introduction could lead to widespread concealment of contamination of food, a lowering of food hygiene standards and an increased risk to public health.

Wholesomeness of irradiated food

In the English language, wholesomeness is a word that combines the ideas of nourishing and healthful. There is no similar word in other languages. This is unfortunate. In the international forums, such as those of the United Nations expert committees, the word has been debased so that it now only covers the concept of *safety* and safety only narrowly defined as the *absence of*:

a) harmful toxic chemicals;
b) harmful microbiological effects;
c) significant impacts on nutrition;
d) induced radioactivity in the food.

Most people will find it reassuring to know that, provided irradiation is properly controlled, the food is not made radioactive. On the other safety aspects the situation is less clear cut. It is possible to argue that the risks are slight, and likely to be acceptable if the whole process is properly controlled. What is not acceptable however is the use of the term 'wholesome' for foods that are to be sold as fresh but, in fact may have been significantly denatured in the process of irradiation. There is a big difference between safe for human consumption and wholesome. This is a distinction the public can and do make and which they need to be allowed to continue to make as a fundamental right.

If consumers are not to be misled we should insist on clear and unambiguous labelling of all irradiated food products.

Vitamins

Irradiation does severe damage to most vitamins, particularly vitamins A,C,D,E & K and some of the B vitamins; B_1 in particular, but also B_2, B_3, B_6, B_{12} and folic acid are affected to some extent.

The extent of the loss depends on the vitamin, on the type of food and the dose given. Fruit juices will suffer more than fresh fruits and these more than vegetables, grains and meat products. Generally speaking the more complex the food the less it suffers vitamin losses during irradiation.

In an attempt to justify the claim that these losses are not nutritionally significant it is said they are:

Table 2: SOME REPORTED PERCENTAGE VITAMIN LOSSES

(Sources: Refs. 1, 2, 12)

Food	Vitamin A	B_1	B_2	B_3	B_6	B_{12}	C	E
Milk	60-70	35 -85	24 -74	33	15 -21	31-33	—	40-60
Butter	51-78							
Cheese	32-47							
Grains and Flour								
Wheat	—	20 -63	—	15	3	—	—	—
Oats		35 -86						7-45
Rice		22						
Beans	—	—	48	—	48	—	—	—
Meats								
Beef	43-76*	42*-84*	8 -17*	—	21*-25*	—	—	—
Pork and Ham	18*	96*	2*	15*	10 -45*	—	—	—
Chicken	53-95*	46 -93*	35*-38*	—	32*-37*	—	—	—
Eggs	—	24 -61*	—	18	—	—	—	17
Fish								
Cod	—	47	2*	—	—	—	—	—
Haddock		70*-90*	4*					
Mackerel		15 -85*			26			
Shrimp	2-27	70*-90*						
Potatoes	—	—	—	—	—	—	28-56	—
Fruits								
Fruit Juices	—	—	—	—	—	—	20-70	—
Nuts	—	—	—	—	—	—	—	19-32

Note: some vitamins are relatively undamaged by radiation but absence of a figure in the table above does not imply that a food has been cleared. In general more work needs to be done on a comprehensive study of vitamin losses.

* All % losses are at doses below the 10kGy proposed clearance level unless *. In these cases doses are below 60 KGy being suggested for sterilisation of meat products.

a) no more serious than occur during normal storage

b) no more serious than occur in cooking and heating.

The first statement is untrue and the second is misleading. Vitamin B_1, for example, is not only very sensitive to radiation damage but the further losses during storage are greater than expected during normal storage of unirradiated foods.[12] Vitamin E is not only virtually destroyed in some irradiation processes, but is destroyed even if it is added as a supplement after irradiation.[1]

As far as cooking is concerned, it is true that vitamin losses, particularly of vitamin C, do occur, but it is the grains, meats and some vegetables that are usually cooked whereas we obtain a large portion of our vitamin C from fresh fruits and uncooked vegetables. The loss of vitamin B_1 during cooking is again greater when the food has been irradiated.[1]

It may well be true that a family with a normal healthy diet in a developed country like Britain will not suffer from major nutritional deficiencies as a result of irradiated food. We do however need to be concerned about those whose diet is already deficient both at home and abroad. This is far more widespread than usually appreciated. Poor eating habits are often forced upon people by poverty, or by the pattern of family working that leaves them overdependent on processed 'convenience' foods.

There has, so far, been no detailed study of the impact that irradiation of a significant portion of our food would have on people on low income diets. Without this study the full impact of irradiated food on diet cannot be said to be satisfactory. Since irradiated food with significant vitamin losses may still look fresh it is extremely important that it is clearly labelled so that the consumer is not misled.

Food quality

The appearance, feel, texture, taste and smell of food are the most immediate perceptions of food quality. Irradiation may inhibit the ripening of fruit. It also produces a softening of the tissues and leaves many fruits more susceptible to bruising. Of 27 fruits investigated, the process has been found beneficial for only eight.

Table 3: RESPONSE OF 27 FRUITS   (Source: Ref. 1)
TO IRRADIATION

BENEFICIAL EFFECTS (a) delay in ripening (b) delay in ageing or (c) control of storage decay.

BANANAS	(a)
MANGOES	(a)
PAPAYAS	(a & b)
SWEET CHERRIES	(b)
APRICOTS	
TOMATOES	(c)
STRAWBERRIES	
FIGS	

NOT BENEFICIAL (d) damaged by radiation (e) accelerated ripening or (f) no positive benefit.

PEARS	(d)	PEACHES	(e)
AVOCADOS		NECTARINES	
LEMONS			
GRAPEFRUIT		PINEAPPLES	(f)
ORANGES		LYCHEES	
TANGERINES		HONEYDEW MELON	
CUCUMBERS			
SUMMER SQUASH			
PEPPERS			
OLIVES			
PLUMS			
APPLES			
TABLE GRAPES			
CANTELOUPES			

In general, fruit suffers more from irradiation than the fungi we wish to kill.

Potatoes may be slower to sprout if irradiated. Once irradiated, the bruising and cuts that occur in harvesting are also slower to heal. For this reason a delay after harvesting is recommended before irradiation. At the same time, irradiation also increases the susceptibility to subsequent fungus attack and rotting. Careful inspection and sorting must remove soil and any potatos that are rotten or damaged. Otherwise the whole batch will quickly spoil. It may be that these factors, brought on by particularly adverse damp harvesting conditions, were reasons for the financial failure of the Canadian Potato Irradiation project in the mid 1960s.[69]

Onions on the other hand are best irradiated within four weeks of harvesting. Damaged or rotting onions also easily spoil the whole batch. Both onions and garlic show signs of internal browning after irradiation,[2] a factor that reduces their commercial value.

Milk and milk products do not irradiate well. Phrases such as 'chalky', 'scorched', 'candle like' or 'burnt wool' have been used to describe the flavour and smell of irradiated milk.[1]

Fats equally suffer from irradiation. 'Musty', 'nutty' or 'oily' have been used to describe the effects. Oils suffer even more. This is of particular importance in attempts to treat oily fish, mackerel for example.[1]

Meat, one of the main foods for which high dose treatments are being developed, develops what has been characterised as a 'wet dog' smell.[1,18]

The exact nature of what has been called these 'typical irradiation flavours' and odours has not yet been fully characterised. Perhaps even more significantly the major culprits in terms of the foods' chemistry have not been fully identified — so little is fully understood about the massive and random rearrangement of the molecular structure of the proteins, fats, carbohydrates, enzymes and residual chemicals in these foodstuffs.

Irradiation and additives

One of the major selling points in favour of food irradiation has been the claim that it will reduce the need for harmful chemical additives in food.[19,20,21,22,23,24]

There is a very real and well founded public concern about the extent to which foods are being adulterated with chemicals. Processed foods rely on additives not merely for preservative effects but for flavours, colours, and bulk fillers too.

In addition there is reason to be concerned about pesticide residues in food and the harmful effects these are having on

Table 4: IRRADIATION AND ADDITIVES

(Source: Ref. 1. Hazards information: Refs. 25, 71)

Some hazardous additives which it is claimed irradiation might replace.		Some additives which might be needed to reduce undesirable effects of radiation.	
*E251 Sodium Nitrite	m. c.	*E251 Sodium Nitrite	m. c.
*E230 Diphenyl	c.	*E230 Diphenyl	c.
E220 Sulphur Dioxide	c.	E221 Sodium Sulphite	i.
E210 Benzoic Acid	a. i.	E300 Ascorbic Acid	t. m.
925 Propylene Glycol	c.	E321 B.H.T.	c.
Chlorine	i.	E320 B.H.A.	c.
E926 Chlorine Dioxide	i.	371 Nicotinic Acide	i. a.
E281 Sodium Propionate	a.	924 Potassium bromate	i.
E236 Formic Acid	i. (banned in UK)	Sodium tripoly-phosphate (TPP)	i.
Ethylene Oxide	c. i. m. t.	Sodium chloride (salt)	
Methyl Bromide	i.	Glutathione (for Vit. B$_1$)	
Ethylene Dibromide	i. c.	Niacin (for Vit. B$_6$)	
Propylene Oxide	i.	Sodium Ascorbate (for Vitamin C)	
Hydrogen Peroxide	i.		

z

Key:
i = irritant, c = carcinogen (known or suspected cancer causing agent in animals and/or humans) m = mutagen (capable of causing mutations), t = teratogen (capable of causing damage to developing fetus), a = capable of causing allergic reactions.

agricultural and other workers. The London Food Commission has produced reports on the health hazards of pesticide residues, [26] and of food additives for both consumers and workers. [25]

It is particularly worrying then to find that, far from eliminating additive use, the process of irradiation will itself require use of a number of additives in order to control some of the undesirable effects of irradiation. [1] Table 4 opposite lists some of the additives that are claimed to be eliminated or reduced by irradiation and those proposed for use with irradiation.

The use of such additives is not to be restricted to the high dose applications where the obnoxious radiation flavours become pronounced but are also proposed for low dose uses to prevent discoloration and other undesirable effects such as bleeding and breakdown of fats in meat. To control these the following steps for preservation treatment of meat are proposed:

1. Cut into portions.
2. Dip in a dilute solution of sodium tripolyphosphate* (or other condensed phosphate).
3. Film wrap.
4. Vacuum pack film-wrapped portions in a bulk container.
5. Refrigerate at 0-5°C.
6. Irradiate with a dose of 1-2 kGy.
7. Ship and store at 0-5°C.
8. Remove from container for display (no more than ½ hour before display).
9. Keep in refrigerated display at 0-5 degrees.
10. Sell within three days.

In this way it is proposed to extend the 'shelf life' of retail cuts of meat to 21 days. [1]

Approval of irradiation, if it is given, should be linked to bans and other restrictions on the use of chemical additives that are

*Sodium tripolyphosphate is a chemical used for cleaning grime off walls (it cuts grease). It is also irritating to the skin and is used as a purgative as well as its use here in food preservation. [27]

currently used as preservatives. Unless this is done we will be adding to the range of hazards not, as those in favour of irradiation claim, reducing the current adulteration of foods.

Toxic chemicals

As noted earlier, bombarding any material with radiation can alter its chemical structure. The first stage of this process is the creation of what are called 'free radicals'. These are highly reactive parts of the original chemical structure that have been split and carry either positive or negative charges. A free radical will rapidly combine with another radical of an opposite charge. In some cases this can lead to recombination of the original molecule. In most cases it will not and a completely new chemical can be created.

These chemicals created in the irradiated food are called 'radiolytic products' or 'radiolytes'. Many of these are similar to chemical changes that occur in other forms of food processing such as cooking. Some, however, are unique to irradiation.[1,2,12]

Because of the complexity of the reactions, it is dificult to identify all of these radiolytic products and to test them in the usually accepted way that chemical additives, for example, would require. Initially, irradiated foods were fed to animals. While, overall, the results have been reassuring, testing of a potential hazard usually involves feeding large quantities of the chemical. Testing by just feeding the food is inadequate. Only small quantities of the unique radiolytes exist in the food. Testing in this way can therefore miss underlying problems or long term hazards.

More recently a number of radiolytic products have been isolated and more normal high dose testing has been done on these.[1] Again the results have been claimed as reassuring. Even though some adverse effects have been found in experimental conditions, it is claimed that these are not likely to occur with irradiated foods under practical conditions, i.e. provided these conform to the international Codex Alimentarius standards and the guidelines developed by the Joint Expert Committee of the WHO/FAO/IAEA.[1]

It needs to be noted that the WHO/FAO/IAEA Committee initially required testing on all irradiated food products. In 1976 this requirement was removed. Results from one food could be applied to another provided that the doses were below a limit of 10 kGy. The US Food and Drug Administration however insists that food irradiated in the medium dose range could contain enough unique radiolytes to warrant toxicological study. All foods irradiated above 1 kGy in the US are required to be tested.

Clearly the dose is a critical factor. The higher the dose the more radiolytic chemicals are created and the greater the potential risk. Initially the WHO/FAO/IAEA specified both maximum and minimum doses that should be used in giving clearance to particular foods.[4] The minimum dose guaranteed that the food would have the changes expected of it, and the maximum guaranteed that undesired effects would be limited. In 1980 this requirement was changed and only an *average* dose was specified. The Committee accepted in doing so that doses up to 50 per cent greater than this average could result.[5] In some quarters this has been interpreted as giving clearance for maximum doses to be 15 kGy.[2]

Thus, though testing has been undertaken over a number of years and the dose ranges within which irradiated food can be regarded as safe have been defined, there has been a steady relaxing of the requirements for testing, and for control of doses. These changes appear to have more to do with commercial considerations than public health.

The public need far more information than simply being told that irradiated food is 'safe'. The public needs to know about the scientific uncertainty that underlies these statements from the expert bodies and to be given details of some of the adverse effects that have been found.

In chemicals that can cause cancer or genetic defects, it is safest to assume that there is no safe level of exposure to such chemicals. Just as with radiation, any dose can cause the initial damage that develops into a cancer. Damage to the genetic blueprint may cause miscarriage or defects in future generations. The fact that a chemical change is small and can be decreased considerably by suitable techniques, does not eliminate the risk: it merely reduces it.[16]

Evidence of Cause for Concern

Lethal effects from feeding irradiated food to mice have been observed. Other studies have not confirmed these effects.[1] The problems surrounding safety in this area are compounded by lack of credibility in some research. In 1968 the American FDA rejected the US Army's research on irradiated pork, and withdrew clearance from irradiated bacon given in 1963. In the 1970s the US Army contracted their research to Industrial Bio-Test Limited. In 1983 the company's officials were convicted of performing fraudulent safety research for industry and government.[29]

Some animals fed irradiated food have been found to have reduced growth, changes in white blood cells and kidney damage. The chemical agents responsible have not been identified but some of the changes may be due to effects on the body's immune system. There is also a suggestion that vitamin reduction in irradiated foods is a cause of some of these health effects.[2]

A similar conflict of evidence exists over whether irradiated food can cause genetic mutations. The balance is in favour of the view that they do not, but some uncertainty remains.[2]

The most recent US scientific research raises questions about safety of irradiated chicken. The report, published in 1984, concludes:

'two of the studies . . . had some possible adverse findings which will require careful consideration before the process can be declared safe'.[32]

Polyploidy — a chromosome defect — has been observed in children, monkeys and rats fed irradiated wheat, and hamsters fed an irradiated diet. Other studies of feeding irradiated wheat or foods irradiated for human consumption to rats have not found polyploidy. It appears that the damage that leads to such chromosome defects decreases when the food is stored and this, together with the variety of components of food, may account for the differences observed. Even so this reinforces the need for vigilance and for setting strict conditions on the process and subsequent storage times for irradiated food products.[30]

As early as 1979 a review of food irradiation literature, for the Hungarian Academy of Sciences, identified 28 reports of adverse effects relating to the radiolytic products created by irradiation in pork. Most of these studies involved much higher doses than those proposed for commercial irradiation. The quantities of a particular radiolytic chemical may increase with dose. The type of radiolytic chemical change is, however, likely to be the same whatever the dose. We are concerned that even small quantities of some chemicals could be hazardous.[106]

Comparison with Other Processes

These uncertainties about safety do not appear in the reports of the WHO/FAO/IAEA or British 'experts' on irradiation. The conclusions are always expressed in terms of the benefits of food irradiation and the safety of irradiated foods. If chemical changes are referred to at all they are said to be 'not significant' or similar to those produced in other processing techniques.

Some radiolytic products *are* similar to those produced by other processing techniques. The quantities produced however can be very different. Hydrogen peroxide, for example, continues to develop after irradiation. Stored foods can have two to three times the levels found in unirradiated foods.[1]

Chemical additives are likely to be needed for use with irradiation. Products created by interaction between these chemicals and radiolytic products need to be tested also. As noted above, some of these additives may present toxic hazards in their own right in any case.[25]

Other common radiolytic products, can be removed after irradiation. This can conflict with the need for sealed packaging to prevent recontamination. Such attempts at removal

'may correct some undesirable consequences of high dose irradiation but leaves other consequences uncorrected and thus the method is of little practical value'.[1]

Concern has also recently been expressed about contamination of foods by some plastic film wrappings.[31] Again it is reasonable to request that studies be done on the changes that occur with irradiation and the combination products of these chemicals with radiolytes when the food is irradiated.

One argument frequently used to dismiss concerns about possible toxic chemical hazards is that irradiation produces effects similar to those found in cooking. In the case of the unique radiolytic products this is clearly untrue. While in general some of the chemical interactions are similar there have been very few studies that directly compare the effects of irradiation and cooking.[1]

It is worth noting that in the case of irradiated meats where doses of 20 to 60 kGy are proposed for sterilisation, a dose of 30 kGy was found to produce some changes that were equivalent to heating meat at 170°C for 24 hours. This is hardly an example of cooking that most people would use for comparison.[1]

Summary of concern about toxicity

Testing of chemical products of irradiation in food indicates that some possibly harmful effects can occur. The way these are dismissed is disturbing. They are frequently called 'insignificant' or 'not likely to occur in practice'. The changes to the rules for testing the hazards, and for limiting the maximum doses to the food is also disturbing. It suggests that there are very strong pressures from interests that want to believe food irradiation is safe and who are more concerned to convince the public than objectively report the problem and uncertainties.

A large number of potentially harmful additives are still approved for use in food in Britain. Any testing system that is less stringent than the one that has allowed these additives to slip through the safety net hardly inspires confidence. For this reason we have stressed that testing of irradiated foods, and the chemical products of irradiation in food, needs to be as strict or stricter than required for additives.

The public have a right to be given unbiased information on the possible harmful effects and not bland assurances that hide uncomfortable evidence under value judgements. It is the job of the public and the political processes to decide whether these effects are significant, and to lay down the regulations under which the risks might be considered acceptable. When science and scientific experts enter this area to pre-empt discussion of these issues we have reason to be concerned.

Microbiological Hazards

There is a possibility of irradiation causing mutations in viruses, insects and bacteria in food leading to more resistant strains.[16] There are numerous examples of insects developing resistance to pesticides for example. So far studies have not shown this to be a major problem with irradiation though some uncertainty remains. Some strains of resistant salmonella have been developed by repeated irradiation under laboratory conditions. Radiation resistant bacteria have been found in environments with high natural or artificial radiation levels[2] and development of such resistance might be a problem around large scale irradiation plants.[1]

Of greater concern, however, is the fact that, though irradiation can kill bacteria on the food, it will not remove the toxins that have been created by the bacteria at an earlier stage.

Increased production of aflatoxins following irradiation was first found in 1973[33] and confirmed in 1976 and 1978.

Aflatoxins are powerful agents for causing liver cancer. The production of aflatoxins was found to be stimulated by irradiation at doses approved by the WHO/IAEA/FAO expert committee.

Table 5: STIMULATION OF AFLATOXIN PRODUCTION   (Source: Ref. 34)

Food	% *increase*
wheat	increases with dose
corn	31
sorghum	81
millet	66
potatoes	74
onions	84

Aflatoxins occur in damp environments on fungus spores on grains or vegetables. Control of humidity in storage becomes even more important in the case of irradiated than non-irradiated foods.

One food where this can be an issue of concern is peanuts. Currently some importers go to considerable lengths to ensure that imported nuts are not contaminated with aflatoxin creating bacteria. They fear that if irradiation is legalised, suppliers may irradiate nuts to keep the bacterial count below the control levels. In doing so, the competitor bacteria will also be destroyed and within a short period after the importers have purchased what they believe to be 'clean' peanuts, aflatoxin production could rapidly increase. It is the processors in this country not the foreign suppliers who will be blamed by the consumer for any subsequent health hazard.

Not all the micro-organisms on food are harmful. Some perform useful functions, particularly in warning us that food is going off by giving off a putrid smell. Yeasts and moulds also compete with harmful bacteria and so provide natural controls on growth of these bacteria.

One example of the possible effects of irradiation arises in considering the use of irradiation to reduce risk from salmonella in chicken and fish.

Irradiation of chicken could kill not only the salmonella bacteria but also the yeasts and moulds that are the natural competitors of botulinum, the bacterium that causes botulism food poisoning. It will also kill the organisms that cause the putrid odour. Yet, at the doses proposed, the botulinum will not be killed. Under the right conditions the botulinum could multiply and become a health hazard without the consumer having any warning smell. [1,35,16]

With fish this is less likely. At the doses proposed there are likely to be enough spoilage organisms left to multiply under similar conditions to the botulinum so that the food smells unacceptable when botulism becomes a hazard. [1]

Clearly such possibilities reinforce the need for strict control of both the irradiation process and the conditions under which irradiated food is stored and handled.

4

Can It Be Controlled?

It is quite legitimate to ask of those who want a new technology 'can it be controlled?' We have therefore been genuinely surprised that those who push food irradiation as a good thing seem to be a bit reticent about their thoughts on controls. The LFC working group therefore addressed the issue of control as a priority. In particular we have been worried about the interests of food workers, the public health monitoring agencies and the consumer's pocket.

Much of the work that goes into putting food into the shops where we buy it, is invisible. Some people act as though food arrives on their plates as if from thin air. Food is work for millions of people in this country. A lot of that work, we know, is done for no wages. We all process food in our homes when cooking. In the formal economy, food is one of the biggest employers. So food irradiation could have a big effect on workers' lives. Talk to any food worker or trades union official and you'll learn of the health and safety problems they already have. Flour means hazardous dust. Packaging on the production line means lots of fast moving objects. Fats in the food mean slippery floors from spillages. So what will food irradiation mean? In this chapter we consider this aspect of safety.

The consumer and food worker in this country have three lines of defence against a raw deal in their food. The law and its sanctions. The public and its vigilance — not accepting inferior foods. And the public health monitoring agencies — the Environmental Health Officers (EHOs), Trading Standards Officers (TSOs) and Public Analysts (PAs). These people are the unsung heroes and heroines of the public health movement. They watch on our behalf what is being done out of sight and before our very eyes. In this chapter we consider the fact that these people have had no training in controlling food irradiation. They have, in fact, no techniques, no tests and a lot of worries and unanswered questions.

Though it is welcome and important that ACINF has recommended labelling food as irradiated, labelling alone will not be enough. We should consider the question of cost. The economics of food irradiation have hardly been discussed by its supporters. Many people in the food trades have told us that they cannot see food irradiation ever being a viable financial proposition. That does not necessarily mean that millions of pounds won't be put into it. So here we consider the existing economics of the British food trades and ask whether food irradiation will be in anyone's interests. Particularly we try to get behind the myth that there is such a person as the 'average consumer' who will benefit from food irradiation.

Working with Radiation

Exposure of food to radiation might, in the end, have some beneficial effects. Exposure of workers has none.

Large doses of radiation can kill by destroying cells in the body so that various organs cease to function, or by damaging the body's immune system leaving it susceptible to disease. Some acute effects such as skin burn, nausea, diarrhoea are also experienced which may not cause death. Years later exposed people may suffer from cancer, or their children from genetic damage as a result of the exposure. Even below doses where any immediate effects are experienced there remains an increased risk of cancer, genetic damage, or susceptibility to disease.[1,11,13,14]

There is no threshold or safe level below which these long term effects do not occur.

When radiation strikes a living cell one of three things can occur:

● any damage done is adequately repaired;

● the cell is killed. In this case, provided too many cells are not killed at once the body will eliminate the dead cells and little harm will be done;

- the cell will be damaged but survive to reproduce in this damaged form. Years later, successive reproductions from the damaged cell may show up as what we call a cancer, or be passed on as a genetic defect to future generations.[13,39]

There is also a growing body of evidence that radiation causes a more general reduction in health by weakening the body's resistance to disease.[14,40]

The crucial point is that there is no dose below which these effects do not occur. It is like walking across a main road blindfold. Do this in the rush hour and you'll be killed. Do it at midnight when there is less traffic and you can be more lucky, but if you do get hit by one of the few vehicles around you can be just as dead. A little bit of radiation does not give you just a little bit of cancer. Any dose, however small, can be the one that does the damage.[13]

The extremely large doses involved in the irradiation of food could result in exposure to workers in the industry. They face considerable risks in the event of malfunctioning equipment, leaking radioactive sources, or accidental exposure to the source. In addition the irradiation chamber will be a very corrosive atmosphere requiring 'regular and preventative maintenance'.[1] Irradiation sources will also need to be produced, transported, stored and installed and the spent sources replaced. This whole process cannot be divorced from the wider processes of the nuclear power programme without which it could not exist. At every stage workers can be, and usually are, exposed to 'low levels' of radiation.[74] Just think what is needed actually to irradiate some food. Workers have got to move highly radioactive substances to plants where food will be irradiated. Other workers will work in those plants. Still other workers will have to service and maintain those plants, and take waste away.

As a general principle any exposure to radiation should be avoided unless it can be justified in terms of some overall benefit. Even then the exposure should be kept as low as possible.[1,41]

In addition there are regulations setting limits on the maximum dose to which workers and/or the public can be exposed. Unfortunately these were set before the principle that there is no safe level was established. As a result, some radiation — using technologies, particularly uranium mining, nuclear fuel handling and plutonium reprocessing, developed under standards that can now been seen as inadequate. It is however very costly both in economic terms, and more importantly in terms of the safety image of these industries to admit the error. Considerable effort, that could be better spent on radiation protection for workers and the public, has gone into attempting to maintain the myth that exposures within the limits are 'acceptable'.

Acceptable Risk?

A person is permitted to receive a maximum of 5 rem per year total exposure to radiation at work, or a maximum of 0.5 rem as a member of the public. A rem is a measure of the biological damage done by radiation. For Gamma radiation, electron beams and X-rays, it is effectively the same as the RAD used to measure doses to food.

> *A new unit, the SIEVERT, has now replaced the rem. Since the Sievert is a large dose compared to the likely worker dose, the milli Sievert (mSv) is commonly used (0.001 sievert) 1 rem equals 10 milli Sieverts.*

Since the 5 rem (50 mSv) limit was set in 1957 it has become clear that exposure to this level of radiation represents a completely unacceptable level of risk.

> *A worker receiving this dose each year would run a risk 8 to 16 times higher than is acceptable for a 'safe' industry, and more than double the risk faced by workers in high risk jobs such as mining.[42,43,44] A 'safe industry' accepts that 1 worker in 10,000 will die each year: or, over a lifetime, 1 in 200 workers will die from an accident at work.*
> *Clearly a risk 8 or more times greater than this is completely unacceptable.*

For a worker to face radiation risks equal to risks in a 'safe industry', the exposure limit will need to be reduced to around 0.5 rem (5 mSv) a year. Even then radiation workers will face all the normal risks of other accidents and so still be doubling their risk of dying from the job.[45]

On top of this, these risks apply only to fatal cancers and serious genetic damage over 2 generations. The non fatal but debilitating effects of other cancers, the less serious or long term genetic effects, and the general lowering of the quality of health from radiation should not be forgotten.

The need for a drastic reduction of the dose limits by at least a factor of 10 is reinforced by the more recent scientific evidence on the risks of radiation exposure.

The national standards and the risk estimates are usually based on the recommendations of the International Commission on Radiological Protection (ICRP) — a self appointed body of experts closely identified with the nuclear industry. [46,13] *Two recent international reviews of the scientific literature have put the risks between two and ten times higher than ICRP* [47,48] *Further evidence suggests that the risks may be higher still.* [49,50,51,52,53]

The most recent evidence from a study of UK Atomic Energy Authority workers indicates that the risk is probably five to six times, and could be 15 times greater than estimates by the ICRP.[53]

New Regulations?

In these circumstances it might have been expected that the UK Health and Safety Executive's new Ionising Radiation Regulations introduced in 1985, would have revised and improved standards for radiation protection. Unfortunately the reverse is the case. In some significant ways these new regulations relax the already inadequate standards currently in force.[54]

Under the new ICRP based system for calculating worker doses from exposure to different parts of the body, the permitted dose to a number of organs is allowed to rise by two to eight times the current limits.

Most of the limits on amounts of particular radioactive materials that a worker can absorb are also relaxed; in some cases quite dramatically.[43] The new regulations do suggest that exposures over 1.5 rem (15 mSv) should be investigated, but at the same time the new regulations only require monitoring of workers in areas where they are likely to receive doses greater than this 1.5 rem investigation level anyway.

Table 6: CURRENT AND PROPOSED LIMITS
FOR ORGAN EXPOSURES   (Source: Ref. 45)

ORGAN	OLD LIMIT (in rem)	NEW LIMIT (in rem)	FACTOR INCREASE
Thyroid	30	50	1.7X
Breast	15	32	2.1X
Bone	30	50	1.7X
Red Marrow	5	42	8.4X
Lung	15	42	2.8X
Gonads	5	20	4.0X
Skin	30	50	1.7X
Extremities	75	50	0.67X
Head (in some circumstances)			3.0X

These relaxations are taking place despite the private view of many within the nuclear establishment that exposures need to be below 1 rem (10 mSv) to be considered acceptable.[56]

Design Limits for Food Irradiation Plants

Clearly in these circumstances food irradiation technology will be introduced into a regulatory framework that does not provide adequate protection for radiation workers. Once established the technology will be yet one more reason why improvements cannot be made in the future. It will undoubtedly then be argued that changes will cost money and jobs.

A number of eminent scientists in the field of radiation research radiological protection have argued for a new limit for radiation technology, 5 mSv, which should be the maximum a worker is likely to receive from all normal operations involving exposure to radiation.[57,58] Unions in the UK and Canada have called for a phased reduction of all radiation exposures to below this limit.[55,59]

For most purposes a 5 mSv limit is already feasible.[42,74] In the case of a new technology such as food irradiation it could be guaranteed, as a plant would be designed to ensure that it was not exceeded.

This will not, as indicated above, remove all risks but would begin to bring these risks into line with those faced in other industries. Just like consumers, workers need to be allowed to make informed choices about the risks and to have adequate protection laid down by a framework of regulations.

Lest these be thought of as peripheral concerns it should be noted that, in 1986, in the USA, the authorities revoked the licence on one plant for 'repeated and wilful' violations of worker health and safety regulations including serious overexposure of some of the workforce.[109]

Regulation and Monitoring of Food Irradiation

There is currently no easy way to detect whether most foods have been irradiated, and if so how many times and with what doses. It may be possible to develop such tests by detecting changes in food chemistry, but these are several years away at least. The only effective control is at the point of processing.[17] Doses, and the conditions such as temperature, packaging, and atmosphere modification all need to be tightly controlled if the undesirable effects identified above are to be avoided.

For these reasons it is reasonable to expect a much greater control over irradiation than is usually required for other food processes. In general irradiation of food is not permitted in Britain. The framework of controls needs to be established before there is any change in this effective ban on food irradiation.

The current regulations would allow an extension of the existing limited permits for medical diets to other food with the requirement only that records are kept of the doses given to the food.[8] This is clearly inadequate. If irradiation is to be permitted at all, a whole new set of regulations will be needed. These should specify maximum and minimum doses and other relevant conditions for each approved foodstuff.

There is also the difficult question of labelling. The 1979 EEC labelling directive requires that the process used to treat a food be included on the label. A reasonable interpretation of this directive is that irradiated food needs to be labelled as such. Since low dose uses will affect foods such as fruit and vegetables that are normally sold as 'fresh' it is vital that the labelling requirement is extended to all irradiated foods and not just packaged foods.

It is worth noting that companion bills have been introduced in both houses of the US congress that would *ban* the labelling of irradiated foods anywhere in the US. These bills also seek to prevent individual states from passing local laws requiring irradiated food to be labelled.[62]

Misleading labelling is almost as bad as no labelling. Symbols like the RADURA, 'emblem of quality' label, used for South African and Dutch irradiated foods, or that proposed in France, should not be considered. The RADURA label may even contravene the Trades Descriptions Act in Britain. Irradiated food should carry in clear and unambiguous terms the label 'IRRADIATED'. Anything less is likely to compound the problems of consumer acceptance by raising the suspicion that the food industry has something to hide.

In view of the uncertainties that exist over the wholesomeness of irradiated foods, particularly at medium and high doses, the blanket clearance for irradiation up to doses of 10 kGy suggested by the WHO/FAO/IAEA would be unwise at this time. If irradiation is permitted at all, clearance should only be given for low dose applications and for specific foods. In each case the maximum and minimum doses and the surrounding conditions should be clearly specified and a certification system established that will follow the food from irradiation to point of sale.

There is an even more difficult problem in the case of food offered for sale through catering outlets. The right to know and make informed choice should apply here also. Caterers, like any retailer or market trader, need to be confident that they know whether their foods either have or have not been irradiated. For consumers to be in the position to choose, they must have confidence in the truth of the labels.

These conditions for permitting irradiation of food should also be applied to all food imported into and traded within the European Community.

As indicated above, there is also the need for companion legislation that would control the use of harmful additives and drastically reduce the maximum permissible doses of radiation to workers.

The impact of any decision to permit irradiation of food extends beyond the concerns of consumers and workers. Siting of an irradiation facility can have consequences for a local authority and community. The very presence of large quantities of radioactive material is a cause for concern. In addition, the transport of the radioactive material to and from the facility needs to be considered. So do planning regulations.

In the USA a plant is currently being investigated after an accident that resulted in radioactive contamination being released into the local sewerage system.[109]

Public consultation and much greater access by locally elected representatives to information on the movement of radioactive material is needed. Strict licensing of the design and operation of irradiation plants also needs to be considered.

As well as the regulations, we must also consider the question of monitoring and enforcement of the legal requirements that are laid down, to ensure that they are being obeyed.

Monitoring of food in the UK is currently undertaken by a combination of Environmental Health Officers, Trading Standards Officers and Public Analysts. These are under control of District and County Authorities. Food regulation is based on the desire to prevent rather than respond to food problems.

Inspection of industrial premises, on the other hand, falls under the Health and Safety Executive's Factory Inspectorate. With irradiation plant the Radiochemical Inspectorate will also have a role to play in monitoring the use of radioactive source material and control of radioactive wastes. Who will be primarily responsible for enforcing the conditions under which irradiation takes place?

It is unclear who is to undertake the difficult monitoring roles and at what levels. It is also unclear what co-ordination and overall supervision will be established. Clearly the public health aspects should be the major concern. If this is to be possible, what training and new facilities are to be made available to Environmental Health officers, Trading Standards Officers and Public Analysts? If other agencies are to be involved how will these be integrated into the monitoring system so that the priorities of protecting the health of consumers and workers is maintained?

The Health and Safety Executive has not only failed to reduce the dose limits for workers, but is in the process of implementing new regulations that would be less stringent than those currently in force. Other bodies associated in the public mind with the nuclear industry do not inspire public confidence. Given the problems that the irradiation of food will face in gaining public acceptance, it is vital that a clear and integrated regulatory system is established that gives priority to the agencies with traditional responsibilities for public health. This needs to be established before the present ban is removed. Prevention is better than curing a crisis.

Economics of Food Irradiation

Food irradiation is likely to be a very costly process, dependent on a variety of factors including: capital costs of the irradiation plant; cost for radiation sources; types and quantities of foods to be irradiated; transport costs; seasonality of food; the market image of irradiated foods, and competition from other processes. The 1981 report of the WHO/FAO/IAEA called for investigation of the economic prospects for food irradiation.[5]

There are some people who will argue that the market place will sort out all these problems. If the price is right the consumer will buy irradiated foods. If not then irradiation won't happen. Reality is more complex than this simple grocer's-shop view of national economics.

An international IAEA Symposium raised many unanswered questions of economics.[123] A full study of the economics of food irradiation should consider the impact the technology would have on the whole food industry, on employment within the industry, and on patterns of marketing and consumption. Just as hidden subsidies can make a process appear financially beneficial, many hidden social costs could indicate that it is after all undesirable. Such a study would help in assessing claims made about the economic benefits of food irradiation.

Costs of an Irradiation Plant

There are currently around 70 pilot or experimental plants operating or under construction worldwide. These vary from portable field units and plants mounted in ships, through to large scale multi-purpose irradiation plants. There are, in addition, a number of medical facilities dealing in sterilisation of medical products attached to hospitals. In Britain there are 10 facilities that currently irradiate medical supplies or animal feed. Of these only four, all owned by ISOTRON plc will be be able to handle commercial food irradiation. One other might be able to do so. Some hospital facilities might be used as part of hospital catering.

Obviously the costs of building such plants vary widely. A large scale production facility based on a US design was costed at £1.25 million in 1984.[64] James Deitch, formerly with the US Department of Commerce, estimated back in 1982 that capital costs for a 3 million Curie Cobalt 60 source plant would then be 1.1 million dollars.[110] A recently commissioned facility for ISOTRON in Britain cost £3.0 million.[61] The US Department of Agriculture has suggested plant costs ranging from 1 million to 11.2 million dollars.[123] The bigger the plant, the greater the economies of scale, the higher the initial costs in its construction. However, the USDA concluded that the economic cost-benefit equation has not yet been determined, as actual costs will be unclear until variable factors such as seasonal use, transport, and consumer acceptability are assessed.

A major running cost in an irradiation plant is the radioactive source. The amount of radioactive source material will depend on the type of food (and hence the dose required) and the amount to be processed. At current market prices a 1 megaCurie Source of cobalt needed for a large facility would cost around £100,000 per year.[64] Certainly, with huge capital and running costs like these, only medium to big food companies will be able to afford their own irradiation facilities.

With costs on this scale and with a virtual monopoly on the current UK facilities, it is only natural to find considerable pressure from ISOTRON for removal of the current ban in the UK. Frank Ley, a Director of ISOTRON was an advisor to the government's advisory committee on irradiated foods. Sir Arnold Burgen, the chairman of the committee, is a part time director of Amersham International, Britain's leading isotope manufacturer.[9,23]

The major source of radioactive cobalt 60 has been Canada where the CANDU reactor is well suited to producing the isotope. Cobalt 60 may soon be produced in large quantities in Britain as radioactive wastes if permission to build the US-style Pressurised Water Reactor is granted. In the meantime the alternative isotope, caesium 137 exists in large quantities as waste from the Sellafield (formerly Windscale) reprocessing plant in Cumbria.[65,66]

In the USA the Department of Energy has been clear that food irradiation is one potential application of nuclear waste material.[124]

Food irradiation plant running costs might come down significantly if cheap radioactive cobalt or caesium sources are provided as a way of dealing with these problem nuclear wastes. Even so, the initial costs are high and there will be considerable pressure to utilize the facilities to the full.

The problem with large scale irradiators is that food needs to be transported from the field to the manufacturer's plant and the irradiated food again transported to storage facilities or markets.

Smaller, transportable 'field' irradiators have been developed for fruit and vegetables, and, on board ships, for fish. Some technical problems such as potato damage in harvesting may limit their usefulness. Maintaining health and safety standards can be a much greater problem in field uses of radioactive materials.

The Manufacturers' Position

The major problem for independent assessment of the economic viability of food irradiation is that there is too little public information made available. Regrettably this is neither surprising nor unusual, as many independent analysts of the food trade know. Much of the Government's information about the content of our food is covered by the Official Secrets Act. The explanation for this state of affairs often given by the Ministry of Agriculture is the need to maintain commercial confidentiality. Many organisations concerned about food secrecy are now calling for the repeal of Section 2 of the Official Secrets Act. International secrets about food irradiation pose still greater problems.

In the USA the Coalition for Food Irradiation was launched in January 1985 with 33 company backers. These included many household food names including General Foods, Heinz, Del Monte Corporation, and Campbell Soup Co. In the UK, the Food and Drink Federation (FDF) is the major umbrella organisation for the food manufacturers. The food industry funded Research Association at Leatherhead has been a focal point for the promotion of the technology in Britain. A Working Party of food manufacturers and irradiation companies has been meeting for several years.

In Spring 1986, just after the Government's ACINF report gave apparent approval to food irradiation, the FDF produced a glossy booklet on food irradiation, as a 'consumer guide'.[111] On the cover was a pretty picture of a child eating a strawberry. Inside there was little acknowledgement of either doubts or concerns about the need and value of irradiation. The message was clear: we think it's good, and want you to as well. In the middle is a picture of some of the foods which may be irradiated — fish, poultry, fruit, vegetables, grains. The FDF booklet gives no information that the nutrients of these foods if irradiated could be seriously affected.

The UK food processors form a powerful and well funded vested interest group. However, they are themselves sandwiched in the food chain between two other very powerful groups, the farmers and retailers.

Britain has a highly concentrated ownership of its food system.[112] Farmland is owned by about two per cent of the population. In food manufacturing, according to one study, there are about 5000 firms but a Government report has suggested that the ten largest companies account for one third of all sales. Approximately two thirds of the trade in some food manufacturing sectors such as oils, fats, biscuits or bread were controlled by the top ten companies according to government figures. In food retailing the top ten

companies in 1985 controlled over half of the market. In 1986 one
of these bought another. That means nine companies control over
half the market. The number of retail outlets nearly halved between
1974 and 1983.[115] Even in catering — a sector famous for its small
businesses — the top ten companies control nearly 60 per cent
of the contract catering market.[112]

This concentrated power over our food is growing rather than
declining. In food and drink manufacturing, the National Economic
Development Office has recently plotted the mergers and
takeovers.[121] There were 22 in 1983, 44 in 1984, and 30 in the first
three quarters of 1985. And that was before the wave of takeovers
which the City of London has dubbed 'merger mania', which
included the biggest takeover of all UK corporate history, by the
Hanson Trust of Imperial Foods, owners of Young's Seafoods.

Clearly, the food magnates have their eyes on important matters.
Big takeovers and battles for market share can lead to astronomic
returns. It has been necessary to remind the Monopolies and
Mergers Commission and the National Economic Development
Office that amidst the merger mania, it is easy to forget food workers
and consumers — and especially those on low incomes.[122]

Many consumers believe that retailers and manufacturers of food
all think alike. They don't. Ever since the 1970s, they have been
fighting an often bitter battle between themselves over prices and
profit margins. Retailers currently hold the sway. With so few retailers
dominating the consumer market what manufacturer can afford
to drop a contract with them? In this context, irradiation can be
seen to offer manufacturers an opportunity to regain the advantage.
If you make or deal in perishable goods, a technology which leaves
the goods looking fresh and fine and which extends their shelf life
could be a bonus. Irradiation also offers food manufacturers a
chance to intervene in some bits of the food chain where they do
not at present, where there is little processing of food. Farmers
and retailers are understandably nervous about being seen to give
consumers what one trade watcher called 'counterfeit food'. It looks
like real food but it doesn't go 'off' or mouldy as it gets old.

Tensions between sections of the food trade are nothing new.
What is new for the UK food business is the likelihood that by the

end of the 1990s there will be no trade barriers between the member countries of the EEC. Or at least that is the wish of the EEC's Vice-President and UK representative Lord Cockfield, who has tabled a Memorandum suggesting that all barriers to internal EEC trade be dismantled.[113] A good thing on first sight perhaps. There could well be uniform regulations about what is and is not acceptable for all European food. This also means that there will be standard rules for irradiation. This would be ideal to the big companies wanting to trade in food across the EEC but has implications for consumers, and for the thousands of food workers across Europe who might lose their jobs.[114] In addition, irradiation could be of benefit to the world food traders. Irradiation can stop sprouting which could be advantageous to the international grain traders.[120]

Consumer Choice — More or Less?

It is clear that there are unlikely to be economic advantages for everyone from the irradiation of foods. Irradiation could give some companies a chance to gain markets in fresh food. 'Shelf life' of vegetables and fruit would be extended to fit into new 'one stop' shopping patterns. Already smaller manufacturers, producers, farmers and retailers are being squeezed out. This process, while allowing some consumers to benefit, reduces the range of choices to many others. However, several large retailers appear to be bowing to customer concerns about irradiation and are making statements that they have no plans to sell irradiated produce.

For consumers with restricted budgets price is often the deciding factor in choosing between apparently identical food products, be it fish fingers or canned fruit. Already there are serious worries amongst doctors and nutritionists about Britain leading the world in coronary heart disease rates.[77,95] Food related illness and ill health generally is heavier among the less well off.[112,118] Discussions with some importers have indicated concern that high quality products could be undercut by lower quality, irradiated food which would have the appearance of quality. Do we want a situation where richer consumers may be able to buy their way out of the 'choice' between irradiated and unirradiated foods, while the less well off are pressured to buy according to price?

Those in favour of irradiation have even implied that the nutrient losses will not matter as it is 'well known' that the British are overeating. This argument is disturbing, as it is untrue. Evidence is emerging in the UK of people under-eating or eating unhealthy diets.[112] People can only eat well if good food is there, affordable, and people are well informed.

Those in favour of irradiation also argue that irradiation will reduce waste. Low dose irradiation of meat, for instance, would not only extend shelf life, it would also allow centralised preparation of meat cuts, eliminating the need to do this at the local shop or supermarket counter. On the whole, concentration of economic power allows companies to rationalise production by shedding labour. The costs of unemployment, as we've seen over the last ten years, are passed on to society as a whole. That's why we call for a detailed study of the economics of food irradiation *before* the ban is lifted. Such a study would have to take account of the changing patterns of marketing and employment in the food trades and balance any apparent direct economic benefits, if any, against the social costs and losses of convenience to *all* consumers, not just those with a good disposable income.

Is Labelling Enough?

Some consumer organisations were asked for advice about irradiation in the early 1980s. Unfortunately they appear to have taken the fairly lax view that as long as the experts say it is safe, labelling is all that consumers will be worried about. Perhaps that is why the ACINF report while saying that there were no problems with irradiation suggested that nonetheless, irradiated foods should be labelled as such. Some people have seen this as a bit patronising and a sop to underinformed consumer organisations close to the food industry. To be fair, of course, consumer and food worker watchdogs are only as good as the resources at their disposal. While consumers are many and companies are few in number, resources are allocated in reverse. All too often consumers are forced to accept industry experts because they have no other source of advice. It is regrettable but understandable therefore that the European consumer body, BEUC, early in the 1980s took a rather narrow view of potential consumer concerns.[116] Since the LFC raised so many issues that needed addressing, the National Consumer Council, the National Federation of Womens' Institutes and the National Housewives' Association to name a few have raised serious doubts.[117] The Consumers' Association has tended to sit on the sidelines on this issue. Even the FDF appeared at one stage to be modifying its position and only asking for clearance of irradiation on a food-by-food basis, the toe-in-the-door approach, perhaps taken more for tactical reasons than a concern for consumer interests.[127]

Is Food Irradiation Necessary?

What is emerging is a picture of a process that, far from providing a solution to problems of food preservation, would be only one more step in a variety of food processing techniques; heating, refrigeration, use of chemical additives, reduction of oxygen and moisture, packaging and hygienic handling. Irradiation actually adds to the complexity of food processing because it creates undesirable changes in the food. Many of the other techniques are necessary to reduce this radiation damage. It is perhaps worth asking the fundamental question of whether irradiation is necessary at all.

The answer to this is by no means clear cut. In the area of low doses there might be a justification for irradiation but even here the overall benefits are by no means obvious.

Many fruits do not irradiate well, however irradiation may produce less damage to the food (and be less harmful to health of workers) than some chemical fumigation treatments for controlling insect infestations. This can be important where quarantine arrangements exist (e.g. in the US to control the Mediterranean fruit fly). Other uses appear to have more to do with the ability of the food industry to stockpile or to deliver an apparently acceptable product for sale when long distance shipments are involved. The benefits to the consumer in this area appear to be marginal, particularly when the effects on wholesomeness and nutritional value are considered.

Irradiation of grains to control infestation may also require control of atmosphere and temperature. Packaging is needed to control reinfestation. Under these circumstances it is hard to see that there will be clear cut economic benefits that could not be derived if these other techniques were introduced alone. By far the greatest food losses occur in the warm, moist and less-developed countries which lack the capital to invest in such storage technology. Food irradiation, with its very high capital costs and complex technology rooted in the nuclear industry, would seem to be the last of a series of measures that should be undertaken to improve the situation in these countries. To propose it seriously as the solution would inevitably mean that control of the food reserves of these countries would come even more under the control of multinational agribusiness than they already are. But then maybe that is precisely why it is being proposed?

In the medium dose range we enter an area where the possibilities of detrimental effects outweighing benefits becomes even greater. The major use in this area is to reduce the number of micro-organisms that cause food spoilage in meats in particular. Reduction, not elimination, is the goal. In the course of reducing salmonella risks, we have seen the possibility of creating other more serious health hazards.

Irradiation does not eliminate the need for refrigeration and packaging. At least one author has questioned whether its use is warranted.

'The doses involved . . . exceed the threshold doses for irradiated flavour development in meats if done above 0°C. Irradiation at sub-freezing temperatures would overcome this. However, in view of the rather large costs that would be involved in such processing, including; freezing, irradiation and thawing, it seems unlikely that a meat processor would undertake it voluntarily. While government regulations could impose radicidation, the risk benefit balance for the consumer would need to be considered. In terms of the costs to the consumer it appears to this author, to be an unwarranted application of irradiation.'[60]

The same author goes on to point out that most parasites are inactivated by the freezing temperatures needed to prevent 'off flavours' in irradiated meats anyway. Further irradiation for this purpose would seem to be unnecessary.

If radicidation (at medium doses) to reduce the level of micro-organisms is of marginal benefit, radappertisation (at high doses) is likely to be even less so. The undesirable effects in terms of food odour, creation of radiolytic chemicals, and changes in microbiological balances will be more pronounced. Refrigeration to very low temperatures (−20 to −80°C), use of additives, packaging to eliminate oxygen and prevent recontamination etc. will all be essential to reduce these effects. While this will allow food to be stored for long periods, the existing methods are already able to do this to a large extent. Since these will still be needed it is unclear where the advantage lies. The application for sterile medical diets is already recognised and permitted. At least one major hospital has discontinued its use — finding it 'of little benefit'. It is nice to be able to experiment with 'space age' food techniques, but the decision to permit wholesale irradiation at these high doses should surely follow clear evidence of need, particularly as the technology introduces a number of distinct problems and hazards.

At the present time high dose applications are not being considered for approval. Sterilisation by irradiation alone appears likely to cause more problems than it will solve.

Irradiation can be used to reduce the heat required for sterilisation just as heat can be used to reduce the amount of radiation needed. It may be that some combination of heat and irradiation treatment could overcome some of the disadvantages of irradiation, though the overall balance of risks and benefits is far from clear.

5

Official Opinion or Scientific Fact?

The previous chapters have outlined some of the questions about safety of irradiated foods — questions over which there is cause for concern, and which need to be adequately answered before any decision to remove the current ban on irradiation of food can be made. They have also raised other issues of concern that need to be addressed as part of the debate about the risks and benefits of irradiation technology.

To what extent have these been answered by the British Government's Advisory Committee Report?

The Advisory Committee Report

On 10 April 1986 the Advisory Committee on Irradiated and Novel Foods (ACINF) published it's report: *The Safety and Wholesomeness of Irradiated Foods.*[81] As widely expected the Committee recommended allowing irradiation up to a dose of 10 kGy. This is in line with the recommendation of the Joint Expert Committee established by the IAEA, and involving the WHO and FAO (JECFI). But it is ten times higher than permitted for general use in the USA.

The ACINF report raises a number of disturbing issues — not least of which is the manner in which its findings are presented to the public. Though it discusses a variety of possible problems that might be of concern from food irradiation, it by and large concluded that in each case there are no special safety problems. Nowhere, however, does it give references to the scientific research that it has used. It is therefore impossible in many instances to determine whether the ACINF's conclusions are supported by scientific evidence, and whether evidence suggesting a cause for concern has been overlooked or omitted.

In this section we outline some outstanding issues of concern that arise out of the ACINF report.

These are:

1. Possible health effects not given adequate consideration:
   - Chromosome defects.
   - Nutritional implications.

2. The necessary regulations and controls to ensure safety if irradiation is introduced:
   - Preventing abuses of irradiation — The Imperial Foods Scandal.
   - Developing tests to detect irradiation of food.

3. The prohibitively short consultation period — three and a half months only given for public comment.

How do the experts answer these concerns?

The Safety of Irradiated Food

The superficial treatment of the issue of polyploidy in the ACINF report highlights the concern we expressed in Chapter 3 over the way the issue of the scientific evidence on safety is being handled. We originally outlined this issue in the LFC Report *Food Irradiation in Britain?*[79]

A copy of the report was sent to the Advisory Committee in September 1985. We hoped thereby to get a definitive answer to the problem from this expert body.

Despite this, the ACINF report[81] appears not to have reviewed some crucial evidence and failed to address the fundamental issue — that a problem may exist with freshly irradiated food, i.e. food that has not been stored for a period after irradiation before being consumed.

The report refers to: '. . . one study, in which irradiated wheat was fed . . . to malnourished children, in whom an increased incidence of polyploid lymphocytes was seen.' The report concludes: 'It was found that the abnormal cells disappeared within a few weeks after withdrawal of irradiated wheat from the diet, and we do not think that this transient phenomenon would have any harmful long term consequences.'[81]

There are to our knowledge not one but four studies, three of which indicate that polyploidy (a chromosome defect) may be caused by consumption of irradiated food.[86,87] There is also at least one study that claims not to have found this problem, though it leaves some questions unanswered and raises other issues of concern.[88]

Polyploidy

Aneuploidy is the presence of an abnormal number of chromosomes in the nucleus of a cell. Normally we all have 46 chromosomes. Polyploidy is an example of aneuploidy in which there are extra complete sets of chromosomes, i.e. human polyploid cells may have two sets (92 chromosomes) or three sets (138 chromosomes).

Polyploidy is rare in human cells but some chromosome defects have been linked with exposure to radiation and to some chemicals, and have been correlated with long term effects, e.g. cancer. Polyploidy can result from a simple failure to separate at cell division and is frequently seen in tumour cells, though this is only one example of the bizarre cell forms present in established tumours.

Polyploid lymphocytes have been found after irradiation and in increasing numbers with age[83] but, as yet, there is no evidence of a direct correlation with cancer incidence.[107]

However, to quote an Indian study:[86]

'the long term health hazard significance of polyploidy seen in the children studied here who had received freshly irradiated wheat, is not clear. On this will depend the answer to the question whether irradiated wheat is safe for human consumption' . . . 'Viewed in the light of these observations [of increased polyploidy in children] it is clear that a cautious approach has to be adopted to the whole question of the mutagenic potential of irradiated wheat.'

A cancer connection?

It is also known that any exposure to radiation (and to chemicals) increases the risk of cancer, though this may take years to develop. If the exposure is to the reproductive system it may also cause some form of genetic damage that may take generations to become apparent. The exact process between the initial damage and the possible (but by no means inevitable) cancer is, as yet, unclear. Scientists do not know what mechanisms come into play but it is widely believed that is damage at the cell level and in particular damage to the chromosomes that *initiates* the process, and that exposure to a secondary agent may be needed to *promote* the cancer. [84]

What is particularly worrying is that the initial effect of irradiation is to create free radicals, i.e. highly reactive charged chemical components created by splitting the more stable complex chemical structures in food. Free radicals are believed to be common cancer 'promoters'. [84,85] That is, they promote the second stage developments that turn the initially damaged cells into malignant (i.e. cancerous) ones. Most of the free radicals created by irradiation will eventually recombine into new chemical forms. However, some remain. One of the tests being developed for detecting irradiation relies on the detection of quite small quantities of free radicals or the chemical changes they produce that remain in the food for some time after irradiation. [1,128]

Whatever the mechanisms involved in the chain of events from cell damage to cancer and/or genetic damage, there are few in the medical and biological sciences who would regard an increase in polyploid cells as a trivial matter. That this chromosome defect has been found in children and animals fed irradiated diets is clearly cause for concern though not of itself proof of a causal link between consumption of such foods and these very serious long term effects. Careful evaluation of all the available scientific evidence is clearly warranted.

The evidence

The study referred to in the Advisory Committee report looked specifically at the problem of feeding *freshly* irradiated wheat to malnourished children.[86] This, and the subsequent animal studies[87] that confirmed the initial findings of increased incidence of polyploidy, do suggest that where the wheat has been stored for some time before being used (usually two to three weeks) there is no detectable increase in the incidence of polyploid cells. It was also found that, as the Advisory Committee report noted, return to a normal diet led to a fairly rapid return to 'normal' numbers of polyploid cells being measured. We cannot however agree with the Committee's unsubstantiated opinion that it is unlikely to have long term consequences. If the Advisory Committee has evidence on this, then it should be indicated. At the present time we simply do not know.

There are other studies that the Committee could have cited whose conclusions offer some reassurance in that they did not find increases in polyploidy. A closer look at the primary research reports however raises more doubts than are laid to rest. One of the reports refers to an eight week feeding trial done in Cambridgeshire, England.[88] One group of rats were fed an irradiated wheat-based diet and the number of polyploid cells compared with another group of rats fed on an unirradiated diet (the control group).

This study did not address the issue of damage from freshly irradiated food. The wheat used had been irradiated two weeks before the start of the study. It was thus ten weeks old at the end. In addition the experimenters (to their credit) reported that on week eight they could not find the feed for the experimental animals and concluded that they must have fed it to the control group. After various changes were made to the study protocol to compensate for this they did not find significant differences between experimental and 'control' animals. However, the results from the two scientists who counted the number of polyploid cells showed no consistency even though they were both investigating cells from the same animals. This indicates some of the problems that this technique has. By the same token, it weakens the strength of the initial Indian findings.

How should we resolve this controversy?

The experts involved in our working group, and others in relevant fields, are agreed that the issue of possible chromosome damage from consumption of irradiated food does give cause for concern. On the basis of the evidence available to us this concern cannot be lightly dismissed.

It would have been helpful if the Advisory Committee with access to considerably greater resources than ours, and who were after all charged with the responsibility for investigating all such issues of safety, had at least addressed the key issue — that of possible effects from freshly irradiated produce.

It is entirely reasonable to expect that, immediately following irradiation, there will be a high incidence of free radicals and some reactive chemicals in the food. Many of these will, with time, combine to form more stable chemicals and some of these may in turn escape from the food. After a time the irradiated food will be (chemically speaking) harder to distinguish from the unirradiated product. Whatever undesirable biological effects it may then have on the health of people consuming it will therefore be considerably reduced.

If there is a problem with freshly irradiated produce this will not of itself invalidate the use of irradiation. What will be needed, however, is a system that guarantees that it is stored for a suitable period before being consumed. A simple requirement for date marking that puts a 'do not sell before' or 'do not consume before' label on it at the time of irradiation may be sufficient. The retailers, trading standards officers, and consumers will thereby be warned. The consumer may also need to be given information about the reasons why and any uncertainties involved in this issue so that she or he can make an informed choice.

Further studies are clearly needed. In the absence of acceptable scientific scrutiny of the possible harmful effects outlined above, other measures such as a labelling, and date marking systems need to be considered.

The studies above refer only to irradiation of grains. We may also need further studies to investigate possible chromosome defects arising from feeding other freshly irradiated produce.

Nutritional Implications of Permitting Irradiation of Food

It is widely recognised that irradiation does affect the various components of food in different ways and that these will have some impact on their nutritional value. We will consider two aspects: vitamin losses, and the impact on essential fats in the diet.

Vitamin losses

According to the Advisory Committee, ascorbic acid (vitamin C), thiamin (vitamin B_1), vitamin A, vitamin E and essential polyunsaturated fatty acids (PUFAs) are all damaged by irradiation to a greater or lesser extent. At the same time the DHSS Panel on Novel Foods, which reported to the Advisory Committee has said that 'little is known about the effects . . . [of] irradiation on folate. Since there are possible problems in the area of public health in relation to the intake of folate this needs further investigation'. Folate is an essential vitamin from the group of B vitamins.

Deficiencies of folate and of other vitamins[96] and of minerals[97] have been linked to the development of neural tube defects such as spina bifida for which the British Isles[98] still have one of the highest rates in the developed world. In addition, studies of elderly people and people entering mental hospitals, for example, have found a significant proportion of people with deficiencies in folate.[99] The Advisory Committee concedes that there are some vitamin losses in irradiated foods but makes little effort to find potential problems. It concludes that these are unlikely to be a hazard because they are within the normal range of losses that occur in storage or cooking.

This ignores several facts:

- That irradiated foods are frequently to be sold as *fresh* or at least looking fresh, and by implication are healthy and wholesome; whereas significant (20-80 per cent) losses in some vitamins can already have occurred.[79]
- Irradiation losses are *additional* to the other normal storage and cooking losses. At a time when, following the NACNE and

COMA reports[95] the public is being encouraged to eat more fresh, whole and nutritious foods, this depletion of nutrients seems extraordinary.

- Some vitamins suffer far greater losses after irradiation than normal. Vitamin B_1 losses in storage are accelerated after irradiation. Vitamin E is almost completely destroyed by irradiation and is further destroyed even when it is subsequently reintroduced as an additive.

- The Government itself has been drawing attention to possible deficiencies in the diets of some people — schoolchildren in particular.[89] People on low incomes may already have diets close to, if not below, the recommended levels for some nutrients. Given that irradiation is considered appropriate for some fresh fruits, and for potatoes, fish and chicken, it is clear that some very staple foods could be affected. Indeed, the DHSS report on schoolchildren's diet was released on the same day and from the same government department as the Advisory Committee Report.

The Advisory Committee's view on these issues is that they are unlikely to be a problem because it is unlikely that all food will be irradiated and therefore people are unlikely to experience significant losses. This is hardly reassuring. The Committee seems to be coming dangerously close to saying irradiated food is OK as long as you don't eat it! We have been unable to find any studies that have actually assessed the impact of irradiation on diets of critical groups in the population. This is worrying, given the obvious fact that few people eat a consistently 'average diet'.

Essential fats

Irradiation is known to reduce the level of some essential nutrients in food. There is growing concern over the effects on a number of essential fatty acids in the diet.

If irradiation of food is permitted, will these nutrient losses be significant? The DHSS Committee on Medical Aspects of Food Policy (COMA) recommended that we reduce our consumption of fats, particularly saturated fats, in order to reduce our high rates of early death and illness from coronary heart disease.[95] The subsequent short fall in energy should be made good by an increased consumption of fruit, vegetables, cereals, beans and pulses. In addition COMA made it clear that a shift in the balance of different fats in the UK food supply would be beneficial: less saturated fats, with a partial replacement by polyunsaturated fats (PUFAs).

We are concerned that the irradiation of foods such as cereal grains, vegetables, fruit, nuts and white fish could result in significant dietary losses of the vitamins and fatty acids affected by the process.

Food irradiation on a large scale is experimental. This is admitted by both the DHSS Panel on Novel Foods and the Advisory Committee, which recommended long term monitoring of irradiated foodstuffs for nutritional damage. In our view, such research should be done *before* the widespread application of irradiation to food, not after. The results of all experiments should be made available for public scrutiny. We are very concerned that those nutritional studies which both the Advisory Committee and the DHSS Panel on Novel Foods examined are not fully referenced in the Government's report. In the absence of hard evidence of nutritional changes, the widespread consumption of irradiated food in Britain or any other country would be an unprecedented human experiment.

Necessary Regulations and Testing Procedures to Control Food Irradiation

As well as deficiencies in reviewing the scientific evidence, the Committee also fails to offer any recommendations on the legislative changes that will be needed to control irradiation and to prevent the kind of abuses that have been found to be occurring already.

At the press conference to launch the Advisory Committee's report, the Chairman, Sir Arnold Burgen, was questioned closely about the revelations that a company in one of Britain's major food groups — Imperial Foods — had used irradiation to conceal bacterial contamination on seafoods that were then sold for public consumption in Britain.[101]

It was revealed that the Imperial Foods Group subsidiary Young's Seafoods had found contamination on at least one shipment of prawns that had been imported into Britain, had shipped these to the Netherlands for irradiation and then illegally imported these same prawns back into Britain to be sold under the 'Admiral' label.[125]

In addition, our investigations and those of a Channel 4 TV programme, '4 What It's Worth'[119] reported that Flying Goose Ltd, acquired by Allied Lyons Group, on more than one occasion has sent prawns to Belgium to be irradiated. These prawns were sent to Sweden where, like Britain, the importation of irradiated foodstuffs is prohibited.

Both Britain and Sweden currently do not permit irradiation or importation of irradiated products. Our information is that the practice of concealing bacterial contamination by use of irradiation is more widespread than the Government admits[91] and that other products could already be on the British market. If the current ban is removed without there being stringent controls in place, such abuses will be even more frequent. An area of abuse would merely be legalised.

The reason for concern goes beyond simply concealing bacterial contamination. While irradiation can reduce the bacterial load on foods — known in the trade as the 'bug count' — it will leave unaffected the toxins generated by the earlier bacterial contamination. These can present a very real public health hazard.

Unfortunately, testing food for the 'bug count' is one of the main methods by which port Health and Trading Standards Officers determine the wholesomeness of food that otherwise looks satisfactory. The use of irradiation makes such bacterial testing systems ineffective.

As our survey has revealed, there are no tests available at the present time that will enable port Health and Trading Standards Officers to detect that food has been irradiated, with what doses, or how many times it has been re-irradiated before it is finally sold. Neither are there readily available tests for measuring the toxins directly if such concealment is suspected.

These are approaches that ought to be developed in the future. At the time of writing one of these involves detection of levels of ortho-tyrosine in meats. Another seeks to identify residual free radicals in hard foods, e.g. in bone or chitin and this may also be useful for other dry wholefoods. Other approaches based on changes in food chemistry, chemi-luminescence, or measuring differentials between bacterial and toxin counts are also under consideration. As yet none of these approaches is developed to the point where a public analyst could use them to detect irradiation. These developments are likely to take three years at least.

The IAEA/FAO/WHO joint expert committee was emphatic that irradiation should only be used to extend the shelf life of otherwise wholesome food and never to be used to make unfit food saleable. Yet this is precisely what is already happening in countries where irradiation is permitted and even occurring despite the law, in countries where it is not[5]

At the DHSS press conference Sir Arnold Burgen admitted that it would have been helpful if the Advisory Committee report had reiterated the IAEA/FAO/WHO recommendations on this point.[90] The Advisory Committee report did not address the issue and made no recommendations for the kind of research that will be needed to develop appropriate tests, the changes that will be needed in the activities of the food monitoring agencies, and the changes in the law that will be needed to stamp out this abuse of irradiation. Perhaps most significantly of all it did not recommend delaying introduction of irradiation until such monitoring, control and enforcement systems are developed.

For the Government, Peggy Fenner, the Parliamentary Secretary to the Minister for Agriculture, when asked to take action to stop these abuses, replied that it was a matter for the Port Authorities and that in any case one documented example did not indicate widespread abuse. How many cases are there of imported foods being irradiated? One wonders how many cases we and other independent investigators must uncover (and provide the documentary evidence for) before the government will take action? We need approaches to the other European Countries for joint initiatives to stamp out these abuses of irradiation.[91]

How to implement labelling regulations?

The Advisory Committee report has recommended that all irradiated foodstuffs be labelled 'irradiated'. This concurs with the results from our survey, in which the vast majority of respondents expressed the opinion that this should be so. However, the issue of labelling is not as simple as the report perhaps implies. The LFC survey reveals that ingredients such as herbs and spices are prime candidates for irradiation — and moreover, that the National Spice Information Bureau is keen to see it introduced. It is therefore entirely reasonable to ask whether every food product containing irradiated ingredients will be so labelled? Manufacturers keen to guarantee no irradiated ingredients are already having difficulties, and have to depend on the integrity of their suppliers.[126]

Fresh fruit and vegetables are seen by the food industry as an obvious choice for the irradiation procedure. Yet a large proportion of the fruit and vegetables sold in this country are sold loose on market stalls. As yet there is no test available to enable a trading standards officer to detect whether or not vendors are properly labelling their produce. The difficulties of labelling loose fruit and vegetables are obvious. Unless these are faced by the Government they will make a mockery of consumer choice. When one turns to the problem of labelling food cooked into meals sold in restaurants, the issue becomes complicated even further.

Consumer groups are united in their call for clear and adequate labelling, ensuring that consumers have a choice about whether or not to buy irradiated food. The Government Committee must give assurances that the problems of labelling, as outlined above, will be properly dealt with.

This will require nutritional and date marking labelling as well as merely honestly labelling food as irradiated. It will involve ensuring that all foods, however they are sold — loose, in bulk, packaged, as ingredients, or through catering outlets — are clearly so labelled.

Consumers rightly identify freshness with health and the nutritional value of foods. Consumer research has indicated that many consumers also want the date of irradiation to be given on the label so that old food can be recognised even if, irradiated, it still looks fresh.[92] This will also allow consumers to identify freshly irradiated foods that they may wish to avoid.

Finally, even the best regulations on labelling will be of limited value unless they can be enforced. Until such time as a test for detecting previously irradiated food is developed, and the monitoring agencies are all trained to use it, there is a strong argument for delaying the introduction of this process.

Conclusions

This chapter has looked briefly at only a few of the more obvious deficiencies in the report of the Government's Expert Committee that was widely expected by many of the respondents of our survey to resolve all the issues surrounding the acceptability of food irradiation. This it has clearly not done.

The London Food Commission believes:

1. That the public scandals over the use of irradiation to conceal contamination of food by major British food groups highlight the need for an effective system for monitoring, testing, regulating and enforcing the strictest standards of food safety before irradiation is introduced.

2. That the scientific tests that will be needed to do this do not currently exist and will probably take three to five years to develop; and that

3. As we show in our pilot survey (Chapter 6), manufacturers, retailers and consumers organisations were by no means convinced of the benefits of irradiation before the Advisory Committee Report, and should be no more likely to be convinced after it.

4. There are apparent contradictions between Government policy on diet and nutrition and the damage that irradiation causes to important nutrients.

We therefore have to ask — why the haste? Why this rush to shorten the process of public consultation, and limit representation to government before the public debate and discussion has even begun? Which interests will be served by a speedy introduction of food irradiation into Britain and, by implication, by a foreclosed public debate?

6

Who Wants It?

Why the indecent haste? We have a situation where a new technology, currently banned, is clearly being abused.

The tests that will be needed to detect irradiation do not exist. The training of the agencies who will have responsibility for controlling it has not yet begun.

There are large areas of uncertainty over issues of safety, a need for systematic review of all the scientific evidence, and possibly a need for additional research.

Those pushing irradiation have used a number of arguments. First they said irradiation would enable a reduction of additives. We showed that in fact there could be an increased use of additives to cover up the effects of irradiation. Then they said irradiation would improve food quality. Until we asked 'whose definition of quality?' And now the argument being used is that irradiation will cut food poisoning. Which begs the question of why there is contamination in the first place and how will there not be contamination even after food has been irradiated? In short, we have the feeling that arguments are being used rather than seriously meant. Who knows what arguments wait in store!

There is as yet no clear evidence that there is a need for irradiation, or that it will not be yet one more way of adding to the cost paid by consumers for even lower quality food.

There is certainly a need to study the potential impact of the destruction of essential nutrients on diets of different groups especially those most vulnerable through poverty.

There is clearly a need to draft regulations to prevent abuses, minimise the potential adverse effects and to guarantee consumer's rights. Labelling of irradiated foods will need more than a symbol or words on packaged products. It will need to include information on nutritional losses, and date marking to indicate the age of products that could continue to look fresh for much longer. Such labelling should also be required for foods with irradiated ingredients, foods sold loose, in bulk, or through catering outlets.

We have much to do and much that must be done before considering a change in the law — why then the haste? Who wants irradiation so badly that they are seeking to curtail this debate?

Selling Food Irradiation?

Early in 1985 a series of articles begun appearing in the media suggesting that food irradiation would be made legal soon.

It was suggested that:

1. It is the food industry, and retailers in particular, who were pushing for food irradiation.

   For example:
   Articles in the Observer (5.5.85), *Sunday Times* (4.8.85) and *Supermarketing* (29.11.85) reported (respectively):

   ● 'The food industry wants irradiation...'
   ● 'The food industry is optimistic that the government will give broad approval for low level irradiation of fruit and vegetables...'
   ● 'Food irradiation will be ushered in by food retailers rather than manufacturers...'

   In addition, the *Meat Trades Journal* of 20.3.86 reported 'that the industry was hoping for Government approval for irradiation of food...'

2. Consumers wanted irradiated food because it would last longer and be free from bacteria — thus making it more 'safe', 'natural' and 'fresh'.

   For example:
   A spokesperson for a Food Industry Research Association said on 'The World Tonight' (BBC Radio 4, 5.6.85): 'the consumer wants

fresh foods, they want long shelf-life foods, they want more natural foods ... this [irradiation] is a way of providing that'.

3. The scientific community was fully convinced that there are no safety problems associated with irradiated food.

Even before publication of the report, the Chairman of the Government's Advisory Committee stated that 'there is no evidence to suggest that irradiation of food is harmful to humans' (*Observer*, 5.5.85).

Prof. Alan Holmes, Director of the food industry-funded Research Association at Leatherhead, said in the *Daily Express* (4.2.85): 'I have a dream that for once the public will take the scientists' word and welcome the process as a great step forward.'

On the other hand, the feedback we had been getting from both consumers and some sections of the food industry did not support this.

The Food Irradiation Pilot Survey

Accordingly, the London Food Commission undertook a pilot survey designed to assess official policy and attitudes within major food groups, large and medium sized retailers and voluntary buying groups, plus trade associations and industry-funded Research Groups. We also surveyed consumers' organisations.

It is not intended to present the results of the survey as the definitive view of the food industry's position on irradiation. The initial sample was intentionally biased by selecting the largest firms in each sector in order to assess the views of leading and respected firms. The sample was taken before the publication of the Advisory Committee report and, as some responses indicate, many in the industry were awaiting this report before forming policy. Indeed, many appear to have expected the Advisory Committee to resolve all issues of concern about irradiation. In addition the response rate to the initial survey was low (around 40 per cent of organisations questioned). Nevertheless, with no other information open to the public, these results are a useful indication of the thinking of various groups at the time. The results also suggest that many of the concerns about food irradiation were shared by different sectors of the industry and consumer organisations.[80]

Pattern of response

We received some information by completed questionnaires. (see Appendix 1) We also received considerable information from letters of reply. Both kind of response to our enquiry were allocated to one of the following categories:

A: Those against irradiation.
B: Those with official policy still undecided — but with reservations about its desirability.
C: Those with no official policy and/or those who do not wish to comment.
D: Those in favour, but also with reservations.
E: Those who give virtually unqualified support to irradiation.

The general patterns of response for each of the sectors surveyed are shown in graphs 1 to 4 of Figure 3.

Analysis of these overall responses reveals only two organisations giving irradiation virtually unqualified support. Ten organisations are expressing outright opposition. Of the remainder, the majority were undecided, frequently awaiting the Advisory Committee report or deferring to the Food and Drink Federation for the food industry position on irradiation. It is worth noting that in addition to those opposed, a further five industry organisations stated or implied that irradiated foods were unlikely to be sold by them.

In all, 31 out of 58 responses expressed some degree of concern about irradiation being introduced at the time.

It would be easy to overstate our case here. We are aware that simply being asked a question can tend to elicit a response that will err on the side of caution. Our points are simply that:

a) it was inaccurate for media reports to present the food industry and consumers as united in welcoming irradiation;
b) there existed a degree of concern about a range of issues; and
c) these concerns needed to be addressed by the Advisory Committee in its report.

Figure 3: GENERAL PATTERN OF RESPONSE BY TYPE OF ORGANISATION

1. Major Food Groups

2. Retailers/Voluntary Buying Groups

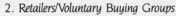

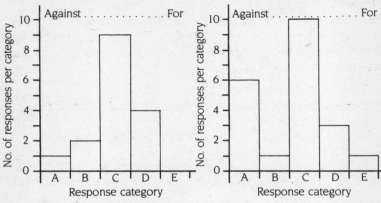

3. Trade Associations/Industry-funded Research Groups

4. Consumer Groups

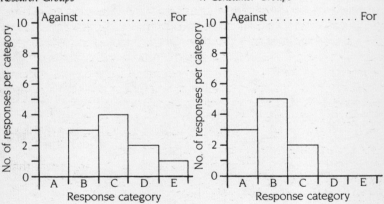

Key to Response Categories

A = Against irradiation.
B = Official policy undecided — but with reservations.
B = No official policy &/or no comment.
D = In favour, with reservations.
E = Virtually unqualified support for irradiation.

In addition to these general positions the survey revealed; an absence of any knowledge about methods of testing for irradiation, and general support for comprehensive labelling of irradiated foods. There was also widespread recognition of a need for additional legislation across a range of issues as shown in Table 7 below:

Table 7: ISSUES REQUIRING LEGISLATION

Area of legislation	% of respondents indicating need for additional legislation
Rank	
1 Labelling	83%
2 Controls on international trade	74%
3 Worker exposure	66%
4 Registration of food premises	59%
5 Date marking of irradiated food	55%
6 Wholesomeness testing	51%
7 Nutritional labelling of foods	47%
8 Improved food hygiene regulations	45%
9 Temperature control regulations	41%

Concern was not confined to consumer organisations.

Many in the industry felt these additional controls would be needed if irradiation were introduced. In addition, many industry organisations and all consumer groups felt there was a need for a public education programme covering the risks and benefits of irradiation before it was introduced.[80]

These findings of widespread concern are reinforced by a survey conducted by the London Borough of Haringey Council of Trading Standards Officers throughout the UK.[108]

It seems from the LFC survey that the food industry was at best divided and frequently undecided about irradiation. While some major firms in the food industry saw some benefits in irradiation, others and especially the retailers and smaller producers had reservations.

Indeed, since the publication of the ACINF report some sectors such as the British Frozen Food Federation have now expressed very strong reservations about the process (Appendix 3). Two leading retailers, Marks & Spencer and Tesco have come out against selling irradiated foods at this time.

Consumers' organisations were even more sceptical, frequently critical, and certainly discriminating enough to see the need for increased regulation and control. They strongly favoured a public education programme that highlights all risks and benefits before it is introduced.

Pressing for Change

While far from being the last word on this issue, for reasons stated earlier, our research does provide some insight into the forces pressing for an early decision to permit irradiation in Britain. These include:

The food industry

Regrettably, one answer to the question 'who wants food irradiation?', could be those suppliers who might wish to sell lower quality and possibly contaminated food to an unsuspecting public. Britain has reason to be proud of the high standards of food hygiene that are maintained in Britain. Outbreaks of food poisoning are rare and frequently traced to lapses in the handling of food. Current practices for cooking, preparation, hygienic handling and refrigerated storage are capable of delivering the vast majority of foods safely to the consumer.

The LFC working group suggests that consumers, the industry and, somewhat belatedly, the Advisory Committee all agree that use of irradiation to conceal contamination of foods is unacceptable. The problem is how to stamp out a practice that is already occurring despite being illegal, and how to prevent the practice spreading if and when irradiation is legalised.

In the absence of the tests that would enable the port Health and Trading Standards Authorities to detect irradiation, it is hard to see any justification for removing the current ban at this time. We are unable to find any clear need for it, and it seems that some of those who want it are those most likely to abuse it.

The pressures for irradiation do not, however, appear to be coming from the food industry alone.

The nuclear industry

Earlier we identified the origins of the process in the 'atoms for peace' programme that was initiated as a cover for the continuing development of nuclear weapons in 1953. We showed how early involvement of the US army in the first ten years' research led to initial approvals for some foods, and how these were revoked by the US Congress in 1968 on grounds of poor scientific research. We identified how the next six years of research data was lost leading to the conviction of the directors of Industrial Bio-Test Limited for conducting fraudulent research for government and industry.

We traced the role of the International Atomic Energy Authority (IAEA) in creating a Joint Expert Committee with the World Health Organisation (WHO) and the Food and Agriculture Organisation of the UN. (FAO). We showed how this committee progressively relaxed the stringency of some testing and control requirements, and expressed concern that its reports lacked references to the scientific literature in support of the conclusions.

Food irradiation technology has its roots not in the food industry but in the nuclear industry. Both the preferred isotopes, Cobalt 60 and Caesium 137 for gamma irradiation are derivatives of nuclear reactor technology. Thirty-six per cent of the IAEA budget goes into research on uses of radiation in food and agriculture — a large portion on convincing the scientific community of the acceptability of food irradiation.

Sir Arnold Burgen, the Chairman of the ACINF, has pointed out that there are limitations on the supply of Cobalt 60.[90] Sir Arnold Burgen is a non-shareholding part-time director of Amersham International, the major isotope manufacturer in Britain. Though currently not involved in production of cobalt for food irradiation, the company has some indirect interests in development of food irradiation now and may have direct interests in the future. It is widely recognised that the major source of the alternative isotope — Caesium 137 — is from radioactive waste generated by nuclear fuel reprocessing at Sellafield in Cumbria.[79] The economic viability of this is questionable — unless there were to be a sudden and rapid increase in demand for caesium.

The whole nuclear industry stands to benefit from any use of radioactive material that can be offered to the public in a positive light.

The food research industry

Earlier we noted a number of reassurances about the safety and benefits of irradiated food. It would not wholly be the case to say in the words of Prof. Alan Holmes on the BBC Pebble Mill at One programme in September 1985, that there are

> 'absolutely no problems whatsoever'.

There are always some scientists, who when given a new technology, would like to be able to rise to the challenge of harnessing it. Such enthusiasm is understandable. However, experience shows that it is essential to be able to monitor and control any new technology. This the industry-funded Leatherhead Food Research Association has at last begun to do research on. It has recently been seeking funding to develop tests for irradiation, and new tests for toxins in foods. It is regrettable that this was not done before they embarked on the enthusiastic promotion of the technology.

The irradiation industry

One group with an obviously direct interest in the removal of the current ban are the companies already involved in irradiation for sterilisation of medical supplies and other materials whose plant could be used for food irradiation. Indeed many public statements on the benefits of food irradiation have come from people with close connections to the irradiation industry. Such statements throughout 1985 have hinted that the UK Government Advisory

Committee was about to give the food irradiation process its approval. A similar pattern emerges from analysis of the press reports from the USA. There, as reported in the press, two major irradiation firms linked the benefits of the process with impending approval by the Food and Drug Administration.

In Britain, a small working group on food irradiation was set up at Leatherhead with representatives of:

- ISOTRON — the company with a virtual monopoly position in the field of gamma irradiation facilities capable of handling food;
- Radiation Dynamics — the leading user of electron beam and X-ray sterilisation techniques;
- Unilever plc — one of the major food companies with strong Dutch connections;
- the Leatherhead Food Industries Research Association.

Two of the leading spokespeople on the advantages of food irradiation quoted in the media have been Alan Holmes of Leatherhead and Frank Ley of ISOTRON.

> Frank Ley has worked in the food research department at Unilever and as principal scientific officer at the United Kingdom Atomic Energy Authority, leading a team investigating the irradiation of food. In 1970 he left to set up a private irradiation company, IPL. In 1984, IPL combined with another company GRS to form ISOTRON.[93]

In September 1983 Frank Ley, the Marketing Director and a leading shareholder in ISOTRON, was appointed as 'economic advisor' to the Advisory Committee.

A number of British MPs have tabled a motion in the House of Commons pointing out their concern over possible conflict of interest. Specifically they noted that predictions of the main recommendation of the Advisory Committee had been widely leaked, not least by Frank Ley, that the company had raised capital through a flotation on the stock exchange, while the committee was sitting, to build a new irradiation plant (ISOTRON already had spare production capacity of 46 per cent at its four existing plants),[94] and that there had been a rise in the capital value of ISOTRON when stories in the financial press linked the future of the company to the impending recommendations of the Advisory Committee. The motion called for an investigation of share dealings in the company (see Appendix 2).[102]

Clearly it does not help public acceptance of the impartiality of the Advisory Committee report to have suggestions of conflict of interest. This would be of little consequence if the scientific evidence put forward by the report was impeccable and verifiable. As we have shown, this is far from the case.

Consumer groups

Given the wide range of issues of concern and the uncertainties remaining after the Advisory Committee report one would expect consumers' organisations to be highly sceptical if not critical of the plans to press ahead with food irradiation. In most cases this is so; however, we note with concern that representatives of the 'WHICH?' magazine-based Consumers' Association have implied that labelling is the only issue of concern to consumers. Since the need for labelling was conceded, albeit reluctantly, by the Advisory Committee out of deference to consumer concerns, it has been suggested that the recommendation that food irradiation is 'safe' should now be accepted in turn by consumers. Other consumer groups have strongly dissociated themselves from such a view. It has to be recognised that an agreement that food should be labelled as 'irradiated' is not enough. Without the means to test for food irradiation there is no way that labelling laws can be enforced. The scandals involving the major food groups noted earlier have shown we cannot rely on the industry to comply voluntarily with the law.

Consumers have the right to know what is in their food and to know it is of the highest standard. As our pilot survey indicated, labelling should clearly state nutrient contents and date of irradiation. All irradiated food must be labelled including foods sold loose, in bulk and through catering outlets.

Some studies into consumer reactions to irradiated foods have suggested strong desires from consumers for food to be labelled with the date of irradiation, so that the food's age could be accurately assessed, without relying on appearance.[92] This could conflict with current UK practice of using 'best before' date marking; another problem ahead for consumers if the ban on food irradiation is lifted.

Who Does Want It?

Our research suggests the following:

- Some sections of the food industry who are already abusing it; and some sections of the food industry who say they may want it to extend shelf life of some foods;
- Some sections of the food research industry;
- The irradiation industry that already has spare capacity and is building even more plants in anticipation of an early decision to legalise it here in Britain;
- The nuclear industry that stands to gain from an extension of the 'atoms for peace' concept into food processing, and perhaps more directly, from growth in isotope production or uses for currently problematic nuclear wastes;
- Some consumers' organisations whose spokespeople seem only to be interested in the issue of labelling as though labelling is the sum of consumer interests at stake;
- The Government, which appears to be turning a blind eye to the abuses that are already happening and to the inadequacy of the scientific advice it is getting.

Above all, the Government seems intent on making an over-speedy decision to approve irradiation of food — curtailing consultation with the public rather than opening up the scientific and public debate.

Who Doesn't Want It?

- Some food importers who see it as undermining food hygiene standards;
- The food and health lobby which is concerned about losses in food's nutritional value and other possible negative health effects;

- Retailers and manufacturers who are worried about being able to give assurances about safety and about public opinion. There are other retailers who oppose irradiation in principle; or who, like Marks & Spencer, argue that it has no place in their policy of selling quality fresh food to consumers on a fast turnover. They simply have no need to extend 'shelf life';
- Many more organisations who, while not opposed to irradiation in principle are increasingly concerned over the scientific uncertainties, the lack of systems for monitoring and controlling the technology, and the way the public debate and consultation process is being handled.

These last include:

Consumers' Organisations.

Major trades unions with members in the food industry.

Trading Standards and Environmental Health Officers.

Environmental groups with traditional concerns over the activities of the nuclear industry.

Independent academics and researchers in the medical and biological fields.

Independent researchers in other fields who have monitored the food industry.

7

Pressure Points

With the kind of vested interests pressing for irradiation it will be necessary for the public to speak clearly and with determination. The problem is where to put the pressure.

Who Will Listen?

The scientific community

Clearly the time for talking with the experts on the Advisory Committee is long gone. Even those who know about the issue or who, like ourselves, had been working on it professionally for some time, found it hard to comment responsibly on the travesty of scholarship that was the ACINF report in the short time allowed.

Nevertheless, within the scientific community there is a code of ethics; a sense that what is done in the name of science needs to be done openly and subjected to independent scrutiny, review, and criticism. Not all of the areas of concern we have identified need, in the final analysis, be grounds for declaring irradiation unsafe. For the present they remain unanswered questions unless one is prepared to accept the answer we have been given, namely; 'We are the experts — trust us it is safe'.

We urge the scientific community to demand a fuller and more open investigation of the issue, and encourage academics and scientists to contribute to the debate on the risks as well as the benefits of irradiation.

Parliament

We have no guarantee that the issue will be debated in Parliament. The current ban relies on the Food (Control of Irradiation) Regulations. These permit exceptions to the general ban and have been used to allow irradiation of special diets for medical patients. It is quite possible to extend permits on a food-by-food basis without a change in the law. This would not require formal Parliamentary approval.

Clearly the Government should be pressed to guarantee a full debate and to put forward additional regulations to control irradiation alongside any proposal to remove the current ban.

More particularly, each MP should be alerted to the concerns in his or her constituency and urged to ask questions of the Government about:

1. Reopening the process of consultation.
2. The lack of openness in presenting the scientific evidence for and against irradiation.
3. The Parliamentary process and timetable for any proposed change in the law and the details of additional regulations to be introduced.
4. Government funding of scientific research into tests for irradiation.
5. Training of local authority monitoring agencies.
6. Government action to prevent abuses of irradiation.
7. Government funding for research into the effects of widespread use of irradiation on diet and nutrition.

MPs not unnaturally, will be influenced by what appears to be an eminent scientific review asserting that irradiation presents no special problems for food safety and wholesomeness. They need to be made aware, by public pressure, that the review by the Advisory Committee is very far from being the last word on the subject and that many people are concerned about unresolved issues that the report has not adequately addressed.

The European Parliament

It is quite possible for the Government to wait out the storm of public protest in Britain, secure in the knowledge that the European Commission is now drafting a directive to 'harmonise' national legislation on food irradiation for the whole of the EEC. If this goes ahead, the decision may be taken out of our hands. It is therefore vitally important that each MEP is alerted to the concern in his or her constituency and that pressure is brought to bear on both the European Parliament and the British Government for a Europe-wide ban on irradiation until these concerns are met.

Consumers' groups

Most consumers' organisations have taken a position which expresses serious concern over irradiation. Some organisations, notably the Consumers' Association, that claim to represent consumers' interests have, however, stated that the only issue of concern is that the food should be labelled. This is clearly not good enough. What will be said on the label? Will this include irradiated ingredients? The European Commission is currently considering a draft directive that would prevent European governments from requiring labelling of irradiated ingredients. What about nutritional labelling and date marking with the date of irradiation so that the consumer will know how old the food is and not be misled by the appearance of freshness?

Even if all of these labelling requirements were met we have to ask; should irradiated food be permitted at all while the concerns over safety, nutrition, and abuse are unresolved. Is it really enough to leave this issue to consumer choice knowing that we are leaving consumers at the mercy of very powerful interests in multinational food companies who have the power to manipulate public perceptions by advertising and their control over the economics of food prices.

Farmers

A major target for consumer pressure will of course be the food industry itself. Already indications are that farmers are worried about the introduction of irradiation, see Farmers' Union of Wales' position

(Appendix 3). With the current food mountains throughout Europe, they are hardly likely to benefit from a system that allows old and lower quality food to be kept even longer.

If we have one recurring nightmare about the use of irradiation it is that it could be used to extend the life of surplus food stocks in developed countries and eventually allow this nutritionally deficient food to be dumped on the 'Third World', adding insult to injury by calling it 'aid'.

Many farmers take pride in providing good quality wholesome fresh food and do not relish the idea of irradiation being used to disguise lower quality. Farmers organisations should be pressed to take a strong position that ensures irradiation is only used where there are real benefits to be achieved by its introduction.

Food companies

Many of Britain's major food companies belong to one of several large food groups that have integrated many different aspects of production and supply under one big, and sometimes, multinational holding company.

The Food and Drink Federation which represents these large food groups has declared itself in favour of irradiation and produced a glossy little booklet extolling its virtues with no mention of any risks or problems.

On the other hand, as our survey has shown many of the smaller and medium-sized companies are more concerned about the risks and the controls that will be needed.

The frozen food industry

The frozen food industry has reasons to fear competition from irradiation. It also believes that with good food hygiene practices in handling and processing, refrigeration is able to to provide safe, hygienic quality food and to provide an adequate extension of shelf life that can benefit both the supplier and the consumer.

Unlike freezers, refrigerators and microwave ovens, you cannot have an irradiator in your kitchen. Equally, if something goes wrong

in your freezer, you will usually know about it and can throw out
problem food. Irradiation technology is large scale and outside
of the control of the consumer.

The British Frozen Food Federation has taken a responsible
position on the issue. Individual companies need to be pressed
to endorse this.

Supermarkets and the retail trade

Pressure on all the food manufacturers and suppliers is important.

Within the food industry, the companies most vulnerable to
consumer pressure are the retailers, particularly the large
supermarkets. Marks & Spencer was the first food supermarketing
company to declare that they had no use for irradiation. They said
that their policy of providing quality food through a system of supply
that ensures a fast turnover, meant that they had no need for
extended shelf life by use of irradiation. Other supermarket chains
are recognising that it may not be in their interests to sell irradiated
foods.

Do other supermarkets have a different policy? If so what is it
and why should this make irradiation attractive — if indeed it is
attractive to them?

All supermarkets should be pressured to declare where they stand
on use of irradiation and in particular asked for guarantees that
they will:

1. Oppose irradiation until tests are available to regulate and control
   it, and until safety considerations have been resolved.
2. Guarantee that all irradiated foods will be labelled including foods
   with irradiated ingredients and foods sold loose or in bulk.
3. That they will insist that irradiated foods are marked with the
   date of irradiation.

The health food industry

A number of health food companies are in fact subsidiaries of big
multinationals. If they are sincere about offering a true healthy food
alternative there should be no doubt about their position on

irradiation. Equally, many of the smaller companies in this sector of the food industry are genuinely trying to offer good quality food at prices people can afford. Either way, consumer pressure should mobilise all these firms to take a clear stand against irradiation at this time.

The companies which oppose irradiation

Some companies within the food industry have already taken a clear position on the issue and have bravely spoken out against the abuses that they know have occurred. These companies need supporting both in terms of buying their products rather than those of companies which have either abused the system, are intending to use irradiation, or who have refused to declare their intentions.

They should also be sent letters of support to show that their stand has been appreciated. There are considerable economic pressures that can be exerted on smaller firms by big food companies. A clear signal that there are benefits to taking a stand as pressure mounts from some quarters for the introduction of irradiation, will help more companies that have reservations about irradiation to join in the call for retaining the ban.

Local Government

Local Government is a major purchaser of food in municipal catering and school meals for example. Other public sector purchasers include the District and Regional Health Authorities. These are democratic bodies and, as such, are amenable to influence by local opinion. They can be asked to adopt a moratorium on any purchasing of irradiated foods until the issues of concern have been adequately resolved. This was first done by the Greater London Council before its abolition, closely followed by the London Boroughs of Haringey and Greenwich in 1986.

Letters to local councillors and to local representatives on Health Authorities can have considerable impact in both alerting elected representatives to the issues of concern that have yet to be resolved, and to the need for a moratorium position.

The nuclear connection

Finally; as we have shown, the major impetus for irradiation has come, not from the food industry, but from the nuclear industry.

Food irradiation is, as Dr Edward Radford, a former Chairman of the US National Academy of Sciences Committee on the Biological Effects of Ionising Radiation has said, 'a technology looking for a use.'

Public pressure

If, after reading this book you are as concerned as we are about the way this need to use the technology appears to be overriding the need for it; if you see it — as we see it — overriding the need for improvements in food hygiene, in food quality, in food safety and the openness of public decision making about food and food policy issues; then we urge you to speak and act now.

We have attempted to provide you with the facts — whether these are listened to now depends on you.

References and Further Reading

1. Edward S. Josephson and Martin S. Peterson (Eds). PRESERV-ATION OF FOOD BY IONIZING RADIATION. (3 vols) CRC Press, Florida USA. Vol 1, (1982), Vol 2 & 3 (1983).
2. P.S. Elias and A.J. Cohen. RECENT ADVANCES IN FOOD IRRADIATION, Elsevier Biomedical Press, 1983.
3. *See* Bibliography in: FAO/IAEA Division of Isotope and Radiation of Atomic Energy for Food and Agricultural Development. TRAINING MANUAL ON FOOD IRRADIATION TECHNOLOGY AND TECHNIQUES. (Second Edition) I.A.E.A. Technical Reports Series No 114 (1982).
4. World Health Organisation. WHOLESOMENESS OF IRRADIATED FOOD. Report of Joint FAO/IAEA/WHO Expert Committee. WHO Technical Report Series No. 604, 1977.
5. World Health Organisation. WHOLESOMENESS OF IRRADIATED FOOD. Report of Joint FAO/IAEA/WHO Expert Committee. WHO Technical Report Series No. 659, 1981.
6. Codex Alimentarius Commission. FAO/WHO, REPORTS OF THE EXECUTIVE COMMITTEE OF THE CODEX ALIMENTARIUS COMMISSION. UNITED NATIONS, 1978 and 1981.
7. *See* answer to Written Question No. 1398/81 by Mr Narjes on 25 November 1982 to Mr Schmid and also answer to Mrs Hanna Walz, Question 1618/83 given 7 February 1984 in the European Parliament.
8. FOOD (CONTROL OF IRRADIATION) (Amendment) REGULATIONS 1972 S.I. 1972, NO. 205, Feb 1972.
9. UK GOVERNMENT'S ADVISORY COMMITTEE ON

IRRADIATED AND NOVEL FOODS. DHSS. Press Release 18, May 1982 for Terms of Reference and Membership.

10. Mr Hansch, Mrs Seibel-Emmerling and Mr Linkhor. MOTION FOR A RESOLUTION ON IRRADIATION OF FOODSTUFFS. European Parliament Working Document 2-1148/84, December 1984. *Also* Mrs Fuillet. MOTION FOR A RESOLUTION — ON THE TECHNIQUE OF IONIZING RADIATION TREATMENT. European Parliament Working Document B 2-550/85, June 1985.

11. John W. Gofman. RADIATION AND HUMAN HEALTH. Sierra Club Books, 1981.

12. P.S. Elias and A.J. Cohen. RADIATION CHEMISTRY OF MAJOR FOOD COMPONENTS. Elsevier Biomedical Press, 1977.

13. *See* T. Webb and R. Collingwood. RADIATION AND HEALTH — A GRAPHIC GUIDE, Camden Press, 1987. *Also* RADIATION ON THE JOB. Tape Slide Presentation for Canadian Union of Public Employees. Radiation and Health Information Service, 1983. *Also*, C. Ryle, J. Garrion, T. Webb. RADIATION YOUR HEALTH AT RISK. Radiation and Health Information Service, 1980.

14. R. Bertell. NO IMMEDIATE DANGER — PROGNOSIS FOR A RADIOACTIVE EARTH. The Women's Press, 1985.

15. J.G. Brennan, J.R. Butters, N.D. Cowell, A.E.V. Lilley. FOOD ENGINEERING OPERATIONS. Applied Science Publishers, 1976.

16. R.S. Hannan. SCIENTIC AND TECHNICAL PROBLEMS INVOLVED IN USING IONIZING RADIATIONS FOR THE PRESERVATION OF FOOD. Dept of Scientific and Industrial Research, Food Investigation Special Report No. 61 H.M.S.O. 1955.

17. FOOD IRRADIATION NOW. Symposum in Ede, Netherlands, by GAMMASTER. 21 Oct 1981. Martinus Nijhof and Dr W. Junk (1982).

18. M.D. Rankin in FOOD INDUSTRIES MANUAL.

19. Robert Millar. COMING SOON ... ATOM RAYS THAT KEEP FOOD FRESH. *Daily Express*, 4 Feb 1985.

20. Sue Thomas. IRRADIATED FOOD. THE FACTS AND FEARS.

Woman magazine, 16 March 1985.

21. Carmen Konopka. NOVEL FOOD STORAGE TECHNIQUES. *Caterer and Hotel Keeper,* 2 May 1985.

22. GAMMA IS GOOD FOR YOU. *The Economist* 22 Jan 1985.

23. Denise Winn. CAN YOU LIVE WITH LONG LIFE FOOD? *Cosmopolitan* magazine, July 1985.

24. Janette Marshall. FOOD THAT DOESN'T GO OFF. *Here's Health* magazine, August 1985.

25. Melanie Miller, DANGER! ADDITIVES AT WORK, London Food Commission (1985).

26. Pete Snell and Kirsty Nicol, PESTICIDE RESIDUES AND FOOD — A CASE FOR REAL CONTROL, London Food Commission, 1986.

27. M. Windholz. Ed. THE MERCK INDEX — AN ENCYCLOPEDIA OF CHEMICALS AND DRUGS. Merck and Co., 1976.

28. POLICY FOR IRRADIATED FOODS, ADVANCE NOTICE OF PROPOSED PROCEDURES FOR THE REGULATION OF IRRADIATED FOODS FOR HUMAN CONSUMPTION. US Federal Register VOL 46, No. 59, 27 March 1981. *See also* FDA Irradiated Foods Committee. RECOMMENDATIONS FOR EVALUATING THE SAFETY OF IRRADIATED FOODS. US Bureau of Foods, FDA, 1980. *And* Code of Federal Regulations Title 21 — Part 179. IRRADIATION IN THE PRODUCTION, PROCESSING, AND HANDLING OF FOOD. US Government Printing Office, 1981. *And* IRRADIATED FOODS — A REPORT BY THE AMERICAN COUNCIL ON SCIENCE AND HEALTH. Oct 1982.

29. K.M. Tucker and R. Alvarez. COMMENTS ON PROPOSED REGULATIONS ON IRRADIATION IN THE PRODUCTION, PROCESSING AND HANDLING OF FOOD. (FDA DOCKET NO. 818-0004). Health and Energy Institute Washington D.C. USA, 1984.

30. D. Anderson and I. Purchase. MUTAGENICITY OF FOOD. In D. Conning and A. Lansdown (Eds). TOXIC HAZARDS OF FOOD. Croom Helm UK (1983).

31. CANCER LINK IN FILM WRAP STUDIED. *The Guardian,* 12 August 1985.

32. D.W. Thayer. SUMMARY OF SUPPORTING DOCUMENTS FOR WHOLESOMENESS STUDIES OF PRE-COOKED (ENZYME INACTIVATED) CHICKEN PRODUCTS IN VACUUM SEALED CONTAINERS EXPOSED TO DOSES OF IONIZING RADIATION SUFFICIENT TO ACHIEVE 'COMMERCIAL STERILITY'. US Dept. of Agriculture, 19 March 1984.

33. Bullerman et al. USE OF GAMMA IRRADIATION TO PREVENT AFLATOXIN PRODUCTION IN BREAD, *Journal of Food Science* 1238, 1973.

34. E. Pryadorshini and P.B. Tuple. EFFECTS OF GRADED DOSES OF GAMMA IRRADIATION ON AFLATOXIN PRODUCTION BY ASPERGILLUS PARASITICUS IN WHEAT. *Cosmet. Toxicology* No. 505, 1979. And AFLATOXIN PRODUCTION ON IRRADIATED FOODS. *Cosmet. Toxicology* No. 293, 1976.

35. FACTORS INFLUENCING THE ECONOMICAL APPLICATION OF FOOD IRRADIATION. Proceedings of a Panel held June 1971. Organised by FAO/IAEA Division of Atomic Energy on Food and Agriculture. IAEA ST1/PUB/331, 1973. Also REQUIREMENTS FOR THE IRRADIATION OF FOOD ON A COMMERCIAL SCALE FAO/IAEA Division of Atomic Energy in Food and Agriculture, 1974.

36. IMPROVEMENT OF FOOD QUALITY BY IRRADIATION. Proceedings of Panel held June 1973. Organised by FAO/IAEA Division of Atomic Energy in Food and Agriculture. IAEA ST1/PUB/370, 1974.

37. OUR DAILY BREAD — WHO MAKES THE DOUGH. British Society for Social Responsibility in Science Agricapital Group, 1978.

38. P.S. Elias. IRRADIATION OF FOOD. *Environmental Health*, Oct 1982.

39. K.Z. Morgan in C. Ryle, J. Garrison & T. Webb. RADIATION YOUR HEALTH AT RISK, Radiation and Health Information Service, 1980.

40. G.V. Dalrymple and M.L. Baker. X RAY EXAMINATION FOR BREAST CANCER : BENEFIT VERSUS RISK. In W.R. Hendee (Ed.) HEALTH EFFECTS OF LOW LEVEL RADIATION Appleton Century Crofts, 1984.

41. ANNALS OF THE INTERNATIONAL COMMISSION ON RADIOLOGICAL PROTECTION — ICRP REPORT 26. Pergamon Press, 1977.

42. US Environmental Protection Agency Office of Radiation Programs PROPOSED FEDERAL RADIATION PROTECTION GUIDANCE FOR OCCUPATIONAL EXPOSURE. EPA 520 4-81-003, Jan 1981.

43. US Nuclear Regulatory Commission. STANDARDS FOR PROTECTION AGAINST RADIATION. -REVISION OF 10 CFR PART 20. Draft [7590- 01] US/NRC, Jan 1983.

44. LIMITATION OF EXPOSURE TO IONIZING RADIATION — EXPLANATORY NOTES RELATING TO A PROPOSED AMENDMENT OF THE ATOMIC ENERGY CONTROL REGULATIONS. Consultative Document C-78 AECB, Canada, 14 Nov 1983.

45. Brief of the Ontario Hydro Employees Union to the Atomic Energy Control Board on Proposed changes to Regulations under the Atomic Energy Control Act. OHEU, Toronto, Canada, Jan 1984.

46. P.A. Green. THE CONTROVERSY OVER LOW DOSE EXPOSURE TO IONISING RADIATIONS. MSc Thesis in Occupational Health and Safety. The University of Aston in Birmingham, Oct 1984.

47. United Nations Scientific Committee on the Effects of Atomic Radiation. SOURCES AND EFFECTS OF IONIZING RADIATION. Report to the General Assembly. United Nations, 1982.

48. Committee on the Biological Effects of Ionizing Radiation (BEIR III). THE EFFECTS ON POPULATIONS OF EXPOSURE TO LOW LEVELS OF IONIZING RADIATION. US National Academy of Sciences, 1980.

49. R. Bertell. HANDBOOK FOR ESTIMATING HEALTH EFFECTS FROM EXPOSURE TO IONIZING RADIATION. Institute of Concern for Public Health, 1984.

50. T. Mancuso, A. Stewart, G. Kneale. RADIATION EXPOSURES OF HANFORD WORKERS DYING FROM CANCER AND OTHER CAUSES. *Health Physics* Vol 33 365-385, 1977. *See also*

British Journal of Medicine, Vol 28 156-166, 1981. And Ambio Vol 9 No. 2.

51. W.E. Loewe and E. Mendelsohn. NEUTRON AND GAMMA DOSES AT HIROSHIMA AND NAGASAKI. Lawrence Livermore Laboratories, 1981.

52. T. Wakabayashi, H. Kato, T. Ikeda, W.J. Schull. STUDIES OF THE MORTALITY OF A BOMB SURVIVORS, REPORT 7-111. INCIDENCE OF CANCER IN 1959-1978 BASED ON THE TUMOUR REGISTRY, NAGASAKI, *Radiation Research*. 93, 112-146, 1983.

53. V. Beral, H. Inskip, P. Fraser, M. Booth, D. Coleman, G. Rose. MORTALITY OF EMPLOYEES OF THE UNITED KINGDOM ATOMIC ENERGY AUTHORITY 1946-1979. *British Medical Journal* Vol 291, Aug 1985.

54. The Health and Safety Commission. IONISING RADIATION REGULATIONS 1985, HMSO, 1985.

55. SUBMISSION OF THE CANADIAN LABOUR CONGRESS TO THE ATOMIC ENERGY CONTROL BOARD ON PROPOSED REVISIONS TO REGULATIONS UNDER THE ATOMIC ENERGY CONTROL ACT. Canadian Labour Congress Ottawa, Canada, Jan 1984.

56. NOTE FOR THE RECORD OF A DISCUSSION WITH DR A.S. McLEAN AND OTHER SENIOR STAFF OF NRPB ON THE IMPLICATIONS OF ICRP PUBLICATION 26, at Risley on 19 July 1978. UKAEA, Sept 1978.

57. B. Lindell. STATEMENT TO PUBLIC FORUM ON NUCLEAR POWER — Middletown P.A. USA, 1983. Also Bo Lindell, D.J. Beninson, F.D. Sowby. INTERNATIONAL RADIATION PROTECTION REGULATIONS: 5 YEARS EXPERIENCE OF ICRP PUBLICATION 26 IAEA International Conference on Nuclear Power Experience, Vienna, Sept 1982. IAEA-CN-42/15.

58. E.P. Radford. STATEMENT CONCERNING PROPOSED FEDERAL RADIATION PROTECTION GUIDANCE FOR OCCUPATIONAL EXPOSURES. Hearings of the US Environmental Protection Agency, Office of Radiation Programmes, 1981.

59. EVIDENCE TO THE SIZEWELL INQUIRY, General, Municipal,

Boilermakers and Allied Trades Union, 1984.
60. W.M. Urbain in Vol 3 of PRESERVATION OF FOOD BY IONIZING RADIATION (Reference 1 above).
61. Terry Garrett. ISOTRON PROFITS EXPECTED TO EXCEED £1 MILLION MARK. *Financial Times*, 1 July 1985.
62. Congressman Sidney Morrison (R-Washington) Senator Slade Gorton (R-Washington). FEDERAL FOOD IRRADIATION DEVELOPMENT AND CONTROL ACT OF 1985. House Resolution H.R. 696 Jan 24 1985. Senate Bill S 288 Jan 24 1985. US Congress 1985.
63. COUNCIL DIRECTIVE OF 18 DECEMBER 1978 ON THE APPROXIMATION OF THE LAWS OF THE MEMBER STATES RELATING TO THE LABELLING, PRESENTATION AND ADVERTISING OF FOODSTUFFS FOR SALE TO THE ULTIMATE CONSUMER. European Commission 79/112/EEC 1979.
64. PRELIMINARY ASSESSMENT DATA FOR AN IRRADIATION SERVICE FACILITY. Simon Food Engineering Ltd, Stockport, UK 1984.
65. *See* British Nuclear Fuels Evidence at the Windscale Inquiry 1977.
66. *See* CEGB evidence to Sizewell Inquiry 1983.
67. F.J. Ley. NEW INTEREST IN THE USE OF IRRADIATION IN THE FOOD INDUSTRY in T.A. Roberts and F.A. Skinner (Eds). FOOD MICROBIOLOGY: ADVANCES AND PROSPECTS. Academic Press, London, 1983.
68. L. Whiting, E. Perkins and F. Kumero in S. Tannenbaum, Ed. NUTRITIONAL AND SAFETY ASPECTS OF FOOD PROCESSING Dekker, New York, 1979.
69. L. Pim. GAMMA IRRADIATION AS A MEANS OF FOOD PRESERVATION IN CANADA. Pollution Probe Foundation, Toronto, 1983.
70. DISCUSSION DOCUMENT ON IRRADIATED FOOD. Prepared for the European Consumer Protection Forum, Berlin, Jan 1983.
71. N.H. Proctor and J.P. Hughes. CHEMICAL HAZARDS IN THE WORKPLACE, Leppencott, 1978.

72. South Manchester Health Authority. APPRAISAL OF CATERING METHODS, 1985.

73. WORLD FOOD REPORT. Monday 15 July 1985.

74. J.S. Hughes, G.C. Roberts. THE RADIATION EXPOSURE OF THE UK POPULATION 1984 REVIEW, The National Radiological Protection Board, NRPB — R173, 1984.

75. Prof. A.W. Holmes, Letter to Editor, *Observer*, 9 May 1985.

76. Frank Ley in IRRADIATION OF FOOD BECOMING ACCEPTED by John Young, *The Times*, 31 Jan 1983.

77. C. Walker and G. Cannon. THE FOOD SCANDAL. Century Publishing, 1985.

78. J. Lewis. FOOD RETAILING IN LONDON. London Food Commission, 1985.

79. Tony Webb. FOOD IRRADIATION IN BRITAIN? London Food Commission, 1985.

80. Tony Webb and Angela Henderson. FOOD IRRADIATION — WHO WANTS IT? London Food Commission, 1986.

81. The UK Government's Advisory Committee on Irradiated and Novel Foods REPORT ON THE SAFETY AND WHOLESOMENESS OF IRRADIATED FOODS, HMSO, 1986.

82. Melanie Miller. DANGER! — ADDITIVES AT WORK, London Food Commission, 1985.

83. For thorough treatment of chromosome damage by radiation, *see* John Gofman, RADIATION AND HUMAN HEALTH — Chapter 3. Pantheon Books, 1983.

84. *See* E. Boyland. TUMOUR INITIATORS, PROMOTERS AND COMPLETE CARCINOGENS, *British Journal of Industrial Medicine*, 47; 716-718, 1984.

85. W.A. Pryor. FREE RADICALS IN BIOLOGY. New York Academic Press, 1977.

86. C. Bhaskaram and G. Sadasivan. EFFECTS OF FEEDING IRRADIATED WHEAT TO MALNOURISHED CHILDREN. Am. *Journal of Clinical Nutrition*, Feb 1975, pp.130-135.

87. Vijayalaxmi and G. Sadasivan, *Int. Journal of Radiation Biology*, 27 (2) p.135, 1975.
   Vijayalaxmi, *Int. Journal of Radiation Biology*, 27 (3) p.283, 1975.

88. Tesh. Davidson, Walker, Palmer, Cozens and Richardson.

STUDIES IN RATS FED A DIET INCORPORATING IRRADIATED WHEAT, International Project in the field of Food Irradiation, Karlsruhe. IFIP — R45.

89. R.W. Wenlock (et al). THE DIETS OF BRITISH SCHOOL CHILDREN, DHSS, 1986.

90. Statement of Sir Arnold Burgen at Press Launch of ACINF Report. 10 April 1986.

91. Letter from Peggy Fenner, Parliamentary Secretary to the Ministry of Agriculture, Fisheries and Food, to the Director of the London Food Commission, 1986.

92. Titlebaum, Dubin and Doyle. WILL CONSUMERS ACCEPT IRRADIATED FOODS. *Journal of Food Safety* 5 (219-228), 1983. And Bruhn, Schultz and Sommer, ATTITUDE CHANGE TOWARD FOOD IRRADIATION AMONG CONVENTIONAL AND ALTERNATIVE CONSUMERS, *Food Technology*, Jan 1986.

93. ISOTRON plc. OFFER FOR SALE BY TENDER OF 3,290,088 ORDINARY SHARES OF 25p EACH AT MINIMUM TENDER PRICE OF 120p PER SHARE. J. Henry Schroder Wagg and Co. 27 June 1985.

94. *Financial Times* 1 July 1985; *Daily Telegraph* 1 July 1985.

95. National Advisory Committee on Nutrition Education, Health Education Council 1983; C.O.M.A. Diet and Cardiovascular Disease, H.M.S.O., 1984.

96. R.W. Smithells et al. FURTHER EXPERIENCE OF VITAMIN SUPPLEMENTATION FOR PREVENTION OF NEURAL TUBE DEFECT RECURRENCES, *Lancet* 1027-1031, 1983.

97. M.H. Soltan and D.M. Jenkins. MATERNAL AND FOETAL PLASMA ZINC CONCENTRATION AND FOETAL ABNORMALITY, *Brit. J. Obstet. Gynaecol*, Vol 89, 56-58, 1982.

98. M. Wynne and A. Wynne. PREVENTION OF HANDICAP AND THE HEALTH OF WOMEN, RKP p.26; World Health Statistics, 1979.

99. E.H. Reynolds and M.I. Botez (eds). FOLIC ACID IN NEUROLOGY, PSYCHIATRY AND INTERNAL MEDICINE Raven Press, New York, 1979.

100. Early Day Motion, IRRADIATED FOODS: SAFETY & WHOLESOMENESS, House of Commons Order Paper, 9 April 1986 (*see* Appendix 2).

101. Early Day Motion, IRRADIATED FOODS : ILLEGAL IMPORTATION, House of Commons Order Paper, 9 April 1986 (*see* Appendix 2).

102. Early Day Motion, IRRADIATED FOODS : CONFLICT OF INTEREST, House of Commons Order Paper, 9 April 1986 (*see* Appendix 2).

103. Frank Cook — letter to Barney Hayhoe, Minister for Health. *And* Michael Meacher — letter to Norman Fowler, Secretary of State for Health and Social Services, *And* Brynmor John — letter to Michael Jopling, Minister for Agriculture, Fisheries and Food. July 1986.

104. K. Tucker, PORK IRRADIATION TO CONTROL TRICHINAE, Health and Energy Institute, Washington D.C., 1986.

105. US Government Accounting Office, THE DEPARTMENT OF THE ARMY'S FOOD IRRADIATION PROGRAM — IS IT WORTH CONTINUING. PSAD - 78 - 146, 29 September 1978.

106. J Barna. COMPILATION OF BIOASSAY DATA ON THE WHOLESOMENESS OF IRRADIATED FOOD ITEMS. IN ACTA ALIMENTARIA (3), 205 1979.

107. *See* Jacobs et al. NATURE Vol 193 p.591, 1962.

108. Survey of Trading Standards Officers by London Borough of Haringey 1986.

109. K. Tucker. PORK IRRADIATION TO CONTROL TRICHINAE, Health & Energy Institute, 1986.

110. James Deitch. ECONOMICS OF FOOD IRRADIATION, 17 CRC Critical Review in Food Science and Nutrition, 17 (4), 307, 1982.

111. Food and Drink Federation. A REVOLUTION IN FOOD PRESERVATION, FDF, 1986.

112. I. Cole-Hamilton & T. Lang. TIGHTENING BELTS, London Food Commission, 1986.

113. Lord Cockfield's memorandum 'COMPLETION OF THE INTERNAL MARKET (COM [85]), 310, and COMMUNITY LEGISLATION ON FOODSTUFFS (COM [85] 603).

114. Bakers, Food and Allied Workers Union, Comments on ACINF, July 1986.

115. Institute of Grocery Distribution, GROCERY DISTRIBUTION, April 1984.

116. BEUC, DISCUSSION DOCUMENT ON IRRADIATED FOOD, prepared for 7th European Consumer Forum, Berlin, Jan. 1983 (AgV/BEUC/127/82).

117. National Consumer Council, National Federation of Women's Institutes, National Housewives' Register submissions to DHSS in response to ACINF, 1986.

118. Marriott, M. and McDowall, M. MORTALITY DECLINE AND WIDENING SOCIAL INEQUALITIES, *Lancet*, pp.274-6, 2 August, 1986.

119. Thames T.V. to Channel 4, 4 WHAT IT'S WORTH, 8 April 1986.

120. Dan Morgan. MERCHANTS OF GRAIN, Weidenfeld and Nicolson, 1979.

121. National Economic Development Office. REVIEW OF THE FOOD AND DRINK MANUFACTURING INDUSTRY, February 1986.

122. LFC letters to NEDO Food and Drink Manufacturing EDC, March 1986, and to the Monopolies and Mergers Commission, February 1986.

123. IAEA, FOOD IRRADIATION PROCESSING, Proceedings of a Symposium held in March 1985, IAEA, Vienna, Sept. 1985.

124. R. Switzer. IRRADIATION OF DRIED FRUITS AND NUTS, in IAEA, Food Irradiation Processing, IAEA, Sept. 1985.

125. WE BROKE LAW WITH GAMMA RAY PRAWNS, SAY FOOD FIRM, *Daily Mail*, 3 March 1986. (*See* Appendix 2.)

126. *e.g.*, Letter from Bejam Freezer Food Centres Ltd., to National Housewives Association, 13 August 1986.

127. Statement to the Seminar on Food Irradiation at Salford University by FIRA spokesperson M. Boyar, September 1986.

128. N.J.F. Dodd, A.J. Swallow and F.J. Ley. USE OF ESR TO IDENTIFY IRRADIATED FOOD, Radiat. Phys. Chem. Vol 26, No. 4, pp.451-453, 1985.

Appendix 1

London Food Commission Pilot Survey on Food Irradiation

We would be grateful for your responses to the following points. You may care to use the questionnaire to circle NO/YES responses and tick other options. We would however appreciate details where relevant and any additional comments you may feel are appropriate.

1. Policy

1.1 Is your organisation in favour of removing the current ban on irradiation of food?   NO   YES

1.2 If the current ban is removed for which food products should irradiation be considered? (Please specify.)

1.3 Should all irradiated food products be labelled as such?
                                                        NO   YES

1.4 Should the consumer be provided with a choice of irradiated and non-irradiated foods in the same product line?
                                                        NO   YES

2. The Current Situation and Irradiated Food

2.1 Are you aware of any problems with imports from or exports to countries that permit irradiation?   NO   YES
If YES, please give details.

2.2 Are you aware of any foods that are or have been introduced into the UK that may have been irradiated?   NO   YES
If YES, please give details.

2.3 Do you have access to equipment or expertise for testing that would enable you to detect whether food has been irradiated?   NO   YES
If YES, could you give details.
If NO would you welcome advice?

3. Introduction of Irradiation

3.1 Do you think it advisable that there should be changes in other regulations to accompany the introduction of irradiation?
                                                                                          NO   YES
If YES which of the following should be considered (please tick)

★ labelling of all irradiated foods
★ date marking regulations
★ nutritional labelling
★ temperature control regulations
★ wholesomeness testing of irradiated foods
★ regulations for worker exposure to ionising radiation
★ controls on international trade
★ The Food Act and Hygiene Regulations
★ registration of food premises
★ other (details)

In particular we would welcome your comments on:

a) the question of whether foods should be labelled as 'irradiated' or whether you prefer the use of the US 'picowaved' or the Dutch 'RADURA' symbol to label irradiated foods;

b) the question of whether food that would fail current hygiene standards at any stage in its handling or processing should be permitted to be irradiated and marketed after irradiation;

c) the question of changes to monitoring and testing procedures to cope with the situation where late irradiation of foodstuffs may remove the bacterial count but not the toxins generated by earlier contamination. As you will be aware most wholesomeness testing currently relies on the bacterial count as an indicator of contamination.

3.2 Do you think that there should be a public education programme before introduction of irradiated food that highlights the risks and the benefits of irradiation?

NO   YES

If YES which of the following should be highlighted (please tick)

★ extension of shelf life of irradiated products
★ chromosomal damage to children fed irradiated wheat products and the consequent need to store some irradiated foods before consumption.
★ reduction of salmonella bacteria with irradiation
★ health hazards that could arise if irradiation is regarded as a panacea for food hygiene problems
★ opinions of expert bodies on safety of irradiated food
★ evidence of adverse effects of irradiation on test animals
★ evidence that shows adverse effects can be discounted
★ remaining areas of doubt and controversy indicating a need to weigh risks and benefits
★ chemical additives that can be replaced by irradiation
★ chemical additives that will be needed to offset adverse effects of irradiation
★ the economic effects of irradiation
★ movements of radioactive materials, disposal of radioactive wastes and irradiation plant safety
★ vitamin losses in irradiated foods
★ aflatoxin production, stimulation and inhibition
★ other benefits or possible problems disadvantages (details)

3.3 Do you think that irradiation can be introduced without further debate and/or regulation?   NO   YES

3.4 Do you think that irradiation of food is undesirable and/or unnecessary and should not be introduced at this time?

NO   YES

4. We welcome any additional or general comments you may wish to make.

We would also be grateful for the following information:

Name of Organisation

Name of contact person for future contact

Telephone Number

Please return the Questionnaire and your comments to: The London Food Commission, P.O. Box 291, London N5 1DU.

Appendix 2

House of Commons Motions on Irradiated Foods

- Conflict of interest

- Illegal importation

- Safety and wholesomeness

Notices of Questions and Motions: 9th April 1986

713 IRRADIATED FOOD: CONFLICT OF INTEREST

Mr Dennis Skinner
Mr Tony Lloyd
Mr Richard Caborn
Mr Brian Sedgemore
Mr Robert Litherland
Mr Ernie Ross

That this House expresses concern about conflict of interest between the work of Her Majesty's Government's Advisory Committee on Irradiated and Novel Foods and the changing position of companies who stand to benefit from its recommendations; notes that the predictions of the main recommendation of the Committee, that the current ban on irradiating food can be lifted, have been widely leaked, not least by Mr Frank Ley, the Committee's technical and economic adviser; notes that Mr Ley is the Marketing Director of ISOTRON, a company which is in a semi-monopolistic position to take advantage of a change in law allowing food to be irradiated in Britain; notes that while the Committee sat, ISOTRON, despite its existing production overcapacity, commissioned a new plant and raised capital through a flotation on the Stock Exchange; notes the rise in ISOTRON's capital value when the financial press linked the future of the company to the impending recommendations of the Advisory Committee; believes that Mr Ley's high public profile, designed to help ISOTRON financially, was incompatible with his rôle as regards the Advisory Committee; calls on all the directors of ISOTRON to give an account of their share interests and charges on them since flotation; and calls on Her Majesty's Government to make a statement and the Stock Exchange to carry out an investigation, so as to satisfy themselves that nothing improper has occurred.

714 IRRADIATED FOOD: ILLEGAL IMPORTATION

Mr Frank Cook
Mr Richard Caborn
Mr Robert Litherland
Mr Ernie Ross
Mr Brian Sedgemore
Mr Tony Lloyd

That this House notes with grave concern the admission of a company from the Imperial Food Group that, having had a consignment of prawns condemned by the Southampton Port Health Authority on the grounds of bacterial contamination, it had the consignment irradiated by Gammaster B.V. in the Netherlands to conceal such bacterial contamination and caused them to be reimported; notes that this practice is illegal under the United Kingdom Food (Control of Irradiation) Regulations; notes too that it is also contrary to the recommendations of the Joint Export Committee of the World Health Organisation Food and Agriculture Organisation of the United Nations and the International Atomic Energy Agency, which bodies demand that irradiation be used solely to extend the shelf life of food otherwise wholesome and never be used to conceal contamination, thereby rendering saleable food that is not truly wholesome; notes further that this instance highlights the inadequacy of existing food monitoring provisions and standards of enforcement procedures, presenting a potential public health hazard; calls on the Imperial Food Group to give to their shareholders, and to the shareholders of United Biscuits, a full explanation of how and why they became involved in such a scandalous practise; requires them to give, too, a public undertaking that such a major misdemeanour will not be repeated; calls upon Her Majesty's Government to alert all port health and trading standards authorities to the possible prospect of such malpractice; and finally urges the Director of Public Prosecutions to investigate these matters fully with a view to initiating appropriate legal proceedings against the offending company.

715 IRRADIATED FOOD: SAFETY AND WHOLESOMENESS

Mr Frank Cook
Mr Robert Litherland
Mr Tony Lloyd
Mr Richard Caborn
Mr Ernie Ross
Mr Brian Sedgemore

Mr Stan Thorne	Miss Joan Maynard	Mr Allen McKay
Mr George Park	Mr Bob Clay	Mr Lewis Carter-Jones
Mr Laurie Pavitt	Mr Sean Hughes	

That this House, in welcoming the publication of the report of the Advisory Committee on Irradiated and Novel Foods, draws to the attention of Her Majesty's Government: (a) the public concern over reports in the scientific literature of lowered birth rates, lower growth rates and kidney damage in some experimental animals fed on irradiated diets and the incidence of polyploidy, a chromosomal defect, in children and animals fed on freshly irradiated wheat, (b) the severe loss of vitamins caused in some foods by irradiation and the possible effect this may have on people who, as a result of low income levels, are on a diet already inadequate, (c) the inadequacy of the current Ionising Radiation Regulatory Requirements for the protection of workers in irradiation plants, (d) the possible abuse of irradiation to conceal bacterial contamination of food, whilst leaving unaffected those toxins that may have been created by bacterial contamination at an earlier stage of handling or processing, (e) the absence of adequate methods or systems for testing for such hazards to public health by the port health and trading standards authorities, the only public bodies charged with responsibility in regard to these matters, and (f) the clear need for consumers to have all irradiated food unmistakeably labelled as such at the point of sale, whether sold loose, packaged, in bulk or through catering outlets for consumption on or off the premises.

Daily Mail, Monday, March 3, 1986

We broke law with gamma ray prawns says food firm

By **STEPHEN LEATHER**, Consumer Affairs Correspondent

ONE of Britain's biggest food groups has sold radiation-treated prawns in contravention of health regulations.

The Imperial Group admits it broke the law last year after a consignment of prawns from Malaysia failed quality-control tests.

Instead of scrapping the seafood, it was shipped to Holland, sterilised with gamma radiation and then brought back and sold under the Admiral label to caterers.

Imperial Group — its products include Courage beer, Ross frozen foods, Golden Wonder and Young's Seafoods — said : 'In January 1985 we bought two containers of Malaysian warm water prawns. They were imported through Southampton, where they were tested by the health authorities. Then they went into public cold store.

'In February we took them out. We didn't have to test them again, but we did. We tested 46 batches. Twenty passed but 26 did not — our standards for these tests are very nigh.

'There was nothing to stop us selling the prawns because they had been passed by the health authorities — but our standards are higher. The decision was taken to send them to Holland for irradiation. Holland is a world leader in this type of treatment.'

A certificate shows the Dutch firm of Gammaster treated three batches of cooked, peeled prawns in April. They were then returned to Britain.

The spokesman said that by the end of May, all the radiation-treated prawns had been sold, mainly to Indian restaurants and caterers.

The Department of Health confirmed that it is illegal to sell irradiated food in Britain and that the trading standards office at the port of entry would be investigating.

Prosecution could result in a fine of up to £2,000 or three months' jail.

Irradiation is widely used to sterilise food in Europe and America — there is no evidence that treated food is unsafe. The treatment allows it to be kept for much longer, which can reduce its nutritional value.

```
CERTIFICATE OF GAMMA IRRADIATION No. 214
------------------------------------------

This is to certify that :

GAMMASTER B.V. - Ede - Holland

has given an irradiation treatment
to the following goods :

CUSTOMER :            Young's Seafoods Ltd., London;

PRODUCT :             Bulk IQF cooked & peeled prawns 300/400

QUANTITY :            1. 1278 x 38 lbs
                      2. 180 x 38 lbs and 572 x 25 lbs

CHARGENUMBER :        1. 850423
                      2. 850423-24/NW5-029

IRRADIATION DATE :    1. 23-04-1985
                      2. 23/24-04-1985

IRRADIATION MODE :    JS 9000 IRRADIATOR

IRRADIATION DOSE :    3 kGy

                                    GAMMASTER B.V.
                                    Ede   - Holland


Acc. Controller :


Acc. Plant Manager :
```

Appendix 3:

Statements of Position

The Food and Drink Federation
Given clearance to use the process, the UK food and drink industry will carefully examine the commercial development of this technology to assess its advantages. Experience in other countries indicates that the main potential benefits are:

- the health and well-being of the population should be improved by the destruction of certain food-borne pathogens

- wastage, especially of ingredients kept in storage prior to processing, could be reduced

- the shelf-life of products could be prolonged

- it could offer an alternative to other preservation methods.

These comments represent the views of the UK food and drink industry as embraced by FDF. We feel unable to respond more fully to your questionnaire which extends into extremely detailed aspects relating to the introduction of food irradiation technology that at this stage it would be premature to address. Nevertheless, I hope that you will find the above comments of interest and help in relation to your survey of current opinion within the UK. If you feel that you would like to discuss the introduction of food irradiation into the UK with us, I would be pleased to hear from you.

H. B. Williams
Deputy Director-General

Ministry of Agriculture, Fisheries and Food

From the Parliamentary Secretary

Dear Mr Lang,

Thank you for your letter of 9 April enclosing a copy of a certificate of irradiation for a consignment of imported prawns.

Enforcement of the Food (Control of Irradiation) Regulations is a matter for local and port health authorities. I know that officials of the Department of Health have discussed this incident with the Dover Port Health Authority, who are investigating and will I am sure take action if they think it appropriate. A single incident of this nature certainly cannot be taken as evidence of widespread evasion of the law and I have no knowledge of any other incidents, which in any case would be a matter for local authorities. Importers are required to comply with our domestic legislation regardless of the country of origin and, if action is warranted, it will be taken by the appropriate authorities here.

I have copied your letter and enclosure to the Department of Health and Social Security.

PEGGY FENNER

Position taken by the British Frozen Food Federation after publication of the ACINF Report

BFFF Policy and Recommendations to MAFF and DHSS

We recommend that:

a) Food irradiation should not be legalised until a simple test, which indicates whether irradiation has been applied, is generally available. We strongly recommend that Government affords funding to a suitably qualified research establishment to ensure such a test can become available before legalisation and that it's simplicity is at 'litmus paper' level.

b) The strictest regulatory monitoring system should be in place before legalisation to ensure that food irradiation is never used to conceal or compensate for poor quality hygiene standards in food processing in the UK or abroad. We feel, therefore, that there could well be a need for European wide legislation to harmonise the national regulations on irradiation of food. We recommend that there should also be European wide controls on the importing of irradiated food from other countries, especially from the third world where, realistically, we must recognise that poor hygiene is common place.

c) On the subject of labelling, we feel that the consumer has a fundamental right to know about the processes that have been used to preserve the food he or she is buying. Hence all irradiated foodstuffs need to be labelled as 'irradiated' in clear and unambiguous terms at the point of sale. This condition should apply to packaged foods, bulk foods, food normally sold loose or unpacked and foods offered for sale in catering outlets.

Farmers Union of Wales
Undeb Amaethwyr Cymru
Irradiation of Food

My Union is opposed to the lifting of the ban on the sale of irradiated food in this country, and we have so informed the Department of Health and Social Security and the Welsh Office. We are not convinced that irradiated food will not pose a health hazard to the consuming public (and possibly to those involved in its preparation and processing), since there do not appear to be any generally applicable chemical or physical tests for identifying irradiated foods, our fear is that these will not be clearly labelled as such, and we are apprehensive therefore that irradiation treatment of food would trigger consumer reaction against purchasing the commodities our farmers and growers produce. We believe that this would also disrupt market flow of processed food products by extending their 'shelf life', and so building up stocks to the detriment of throughput of untreated products. We believe that there has been insufficient research into the long-term effects of irradiation, and that more attention should be paid to this aspect of the matter.

E. LEWIS,
General Secretary.

Appendix 4:

Contacts and Addresses

Contacts

The following organisations and contacts may be useful to you in seeking information or action on issues that concern you about food irradiation.

LONDON FOOD COMMISSION,
P.O. Box 291, London, N5 1DU

Sir Arnold Burgen
Chairman, Advisory Committee on Irradiated and Novel Foods, c/o DHSS, Alexander Fleming House, Elephant and Castle, London, SE1

The Rt. Hon. M. Jopling
Minister of State for Agriculture Fisheries and Food, Whitehall Place, London, SW1.

The Rt. Hon. Antony Newton
Minister of State for Health DHSS, Alexander Fleming House, Elephant and Castle, London, SE1

Your Member of Parliament, House of Commons, London, SW1

Your Member of the European Parliament, 2 Queen Anne's Gate, London, SW1

Food Manufacturers

The Food and Drink Federation,
6 Catherine Street, London, WC2

Unilever PLC, Unilever House, Blackfriars, London, EC4P 4BQ

Imperial Foods Ltd,
Clifton House, Goldington Road, Bedford, MK40 3NF

Associated British Foods PLC,
68 Knightsbridge, London, SW1X 7LR

Allied Lyons, Allied House, St John Street, London, EC1

Dalgety Spillers, 19 Hanover Square, London, W1R 9DA

Food Retailers

J Sainsbury PLC,
Stamford House, Stamford Street, London, SE1 9LL

Safeway Food Stores Ltd, Beddow Way, Aylesford, Kent, ME20 7AT

Tesco Stores Ltd, Tesco House, Delamare Road,
Cheshunt Waltham Cross, Herts, EN8 9SL (see Appendix 3)

Waitrose Ltd, 4 Old Cavendish Street, London, W1A 1EX

Asda PLC, Asda House, Britannia Road, Morley, Leeds, LS27 OBT

Argyll PLC, Millington Road, Hayes, Middlx, UB3 4AY

Marks & Spencer PLC,
Michael House, Baker Street, London, W1A 1DN (see Appendix 3)

Spar, 32/40 Headstone Drive, Wealdstone, Harrow, Middx, HA3 5QT

Mace Line Marketing Ltd,
Gerrards House, Station Road, Gerrards Cross, Bucks, SL9 8HW

Co-operative Wholesale Society, New Century House,
P.O. Box 53, Corporation Street, Manchester, M6 4ES

The Dee Corporation, 418 Silbury Boulevard,
Milton Keynes, MK9 2NB

Other Organisations

The Consumers' Association, 14 Buckingham Street, London, WC1

Consumers in the European Community,
24 Tufton Street, London, SW1

National Housewives Association, c/o Flat 4, 68 Somerset Place,
Stoke, Plymouth, Devon

National Consumers Council, 18 Queen Anne's Gate, London SW1

National Federation of Women's Institutes,
39 Eccleston Street, London SW1.

The Bakers, Food and Allied Workers Union,
Stanborough House, Great North Road, Stanborough,
Welwyn Garden City, Herts, AL8 7JA

The General, Municipal, Boilermakers and Allied Trades Union,
Thorne House, Ruxley Ridge, Claygate, Esher, Surrey, KT10 0TL

The Transport and General Workers Union,
Transport House, Smith Square, London, SW1

National Union of Public Employees, 18 Grand Depot Road,
Woolwich, London SE18

The Union of Shop Distributive and Allied Workers,
188 Wilmslow Road, Manchester, M14 6LJ

The Radiation and Health Information Service,
P.O. Box 805, London SE15 4LA

Radiobiology Dept, St Bartholomew's Hospital Medical School,
Charterhouse Square, London, EC1

Food Policy Unit, Manchester Polytechnic, Hollings Faculty,
Old Hall Lane, Manchester, M14 6MR

Institute of Environmental Health Officers,
Chadwick House, Rushworth Street, London, SE1 0QT

London Borough of Haringey, Trading Standards Office,
CPD, 590 Seven Sisters Road, London N15 6HR

Southwark Public Analyst, Municipal Offices
Larcom Street, London SE17 1RY

Friends of the Earth, 377 City Road, London, EC1V 1NA

Greenpeace, 36 Graham Street, London, N1

ISOTRON plc, Moray Road, Elgin Industrial Estate,
Swindon, Wilts, SN2 6DU

Food Industries Research Association, Randalls Road,
Leatherhead, Surrey, KT22 7RY

Index